the dionysia

Propagate: Fruits from the Garden is a theatrical presentation by *the dionysian public library*, graciously printed in association with Western Michigan University's Office for Sustainability, first published on Midsummer 2024. This book is of the earth and from the earth and as such the notion of it being copyrighted is a blasphemous perversion of Gaia's infinite bounty. All of the included works are here reproduced with the permission of the authors, and all rights to the work remain wholly their own except for the right to reproduce the work here. This book was printed on occupied land. Tear down every wall and tend the soil until she's healed.

"Grow" is copyrighted to Jo Rodriguez, 2018, and appears here with the permission of the artist.

This work contains poetry, and is therefore highly volatile and should be handled with care.

Bella ciao.

For Kelsey, who makes this bizarre world so worth fighting for, and for all of us.

*"It never really hurts to know
That there's always another season
Waiting after the fall
And as the springs and the summers come
You'll feel yourself begin to grow"*

—Mystery Girl
"Grow"

Table of Contents

Preface: The Art of Sustainability by Ju Collins................1
An Invitation..6
What I Ask of the Future by Tricia Knoll.....................10
Two Poems by Kostandi...12
Before the Storm by Paul Hilding..................................16
Two Poems by Claire Thyne..18
Brassy by Keiraj M. Gillis..21
Four Poems by Diane Funston..23
False Spring by J. J. Stewart...30
Three Poems by William Doreski...................................34
Orchard Language by Kathy Pon....................................40
Three True Stories by Cassandra Caverhill..................42
Another Revolution by Robert Witmer........................46
Seven Poems by Svetlana Litvinchuk............................47
Poem on a Hummingbird Sage by John Winfield Hoppin..58
JESUS COMING SOON by Katie Bausler.................60
Two Poems by Irena Kaçi...65
In Between by Lawrence Winkler...................................68
Land of Spiders by Kiki Adams.......................................74
Wuthering by J.J. Carey..77
Dear Jody by Karen Pierce Gonzalez..............................78
Two Poems by Kushal Poddar..80
We'll Never Have Paris by Zander Lyvers....................82
Two Poems by Gurupreet Khalsa...................................91
Shore Birds by Phyllis Green...95
Three Poems by Emery Pearson.....................................100
Χαρούπι by Ivars Balkits..104
Jonestown and All the Rest of— by Lisa Lahey............107
Two Stories by Lori Litchman..109
My Sanctuary by Claudia Althoen..................................113
Fallen Fruit by Didi Aphra...115
Natural Extracts by Skye Rozario Steinhagen.............116
Mariana's Headstone by Zach Murphy..........................121
Two Poems by Daniel P. Stokes......................................123
A Few Final Words by Emily Bright..............................127
A Parting...13

Preface:
The Art of Sustainability

by Ju Collins

"No doubt that a small group of thoughtful committed individuals can change the world. In fact, it's the only thing that ever has."

—Margaret Mead

Welcome, friends!

It's been my pleasure to bring my background in all things nature (jack of all trades, Master of Science in Environmental Policy and Management; thanks University of Denver!) to this anthology as guest editor, and therefore to you all. In a time of political, environmental and physical unrest, for the love of Gaia keep writing. It matters, you matter.

Art, not just written but in all forms, brings the subject to eye level for the reader. In our case, this anthology of written work surrounding the relationship between humans, the environment and everything in between them, allows for myriad experiences, perspectives and thought processes to come to life. The first step to stewardship is to be able to intimately know and care about what is happening around us. Without awareness, there is no thoughtfulness, and without thoughtfulness there is no change. To have had an experience that touches your soul or to read a story that resonates and invigorates you into action is what can make all the difference.

In being able to see your surroundings from a different perspective you allow yourself to interact with it in an entirely new or different way. This is why we felt it was so imperative to get these pieces out into the world as a cohesive collection. From a purely scientific point of view, sustainability is entirely necessary for our own survival. Our own sustenance revolves solely around what we sow and the reciprocal relationship we have with the Earth. From a more artistic perspective, preserving nature is *everything*. What a sad world it would be without the beauty and the chaos of life here on Earth. How could we deny sustainability for the Earth itself yet expect asylum and continuity for ourselves? Are we not *of* the Earth? Created, sustained, and diminished back to her, from beginning to end? In my eyes, appreciation is what leads to fierce protection. Having creative work such as that included here as a reminder of what we love most can be the spark that drives us to take the extra step, sign the petition, join the rally; do whatever lights your fire.

Scientifically, there are so many definitions and parameters of "sustainability" that it's enough to make your head spin. I prefer to approach the concept from the heart; to establish *why* we might care to change the way we interact with our environment, how we consume it and how we might sustain it long term. Ultimately, I hope that your reading experience here will further expand your curiosity into how we live and what we long for in our collective future, and how we might take steps towards that as we move through time together. If you feel so inclined after reading, you might find our good friend Google has some excellent suggestions about personal choices (big and small) that have a significant impact towards sustainable living. Every bit counts, and I do mean that sincerely. Every. Little. Bit.

We now live in an age of nearly constant change and "improvement"—depending on who you think the improvement serves and how you managed to attain it. What did it cost? I don't mean how much money, although that seems to be a sticking point for many these days. What did it cost the land? The sea? What was the emotional loss? What happened to all that existed within the parameters of what is now something that serves us, and only us? What did it remove or replace or decimate in order to exist as it does now? There is a theory in ecological science that refers to the biological cost of a project. Maybe that refers to a new apartment complex, the drying of a wetland to build a parking lot, the changing of rules around how a body of water is used recreationally, or the removal of a certain species from an area through hunting or other means. The biological cost is what it will cost us to artificially perform the ecological services of what was once there, the services that we are now removing to build something that is useful to us (such as those mentioned above). The natural process, for example the passive filtering of water through a wetland, still needs to happen in order to fulfill the proper water cycle we rely on for fresh water sources. But now we've dried out the area that performed it naturally for us for free. We must absorb that responsibility (and financial cost) somehow or else we risk losing even more of a precious resource that we already scrounge and wring the Earth dry for. So in response, we build a water processing plant which takes up even more land and resources and creates a positive feedback loop of chasing our own tails. There are truly endless examples of this same loop, much too many to list here. It's a one-way ticket to irreversible climate change, no refunds or exchanges. One research study estimates that beavers (yes, beavers) provide such staggering freshwater reservoir containment and regulation services through building of dams and removal of invasive woody plant species that it would cost us upwards of 200 million dollars (USD) annually to replace those same services if they were eradicated*. That's just some food for thought about a species which is generally considered a "pest" and are often removed from properties if noticed. The concept of

*Thompson, Stella, Mia Vehkaoja, Jani Pellikka, and Petri Nummi. "Ecosystem Services Provided by Beavers Castor Spp." *Mammal Review* 51, no. 1 (2021): 25–39. https://doi.org/10.1111/mam.12220.

sustainability suggests we should put more forethought into how our actions affect our long term survival, not just how it benefits us in the coming weeks, months, or even years. As you read, the hope is that the feelings that arise and the memories that come rushing back will instill a desire to ponder how your own actions might affect how you experience those very same things in the future. How your children might experience them, or even your grandchildren. Sustainability is more than recycling, composting, reducing and thoughtful product choices, even though they're all very important parts of it; its real power is in holding space for the potential for generational experiences in grounding, connection and participation in human relationships and what happens between them and the environment they share.

 Each of us as humans, to be grouped together only by our sameness, has a very unique, dynamic, and emotional connection to nature. Surely there are variations in what you prefer to spend your time connecting with; some people are drawn to the open skies and the birds who fly so freely above, some to the ocean and its moody tides and deep, dark secrets. Maybe it's dogs, cats and other household pets; maybe it's wild game animals that roam the woods near your home that sometimes tread a little too close to the new freeway now connecting here to somewhere. I can guarantee that whatever moves you is connected to what moves the next person, and the next, and so on. We cannot have one without the many, many others. It stands for all, it means not one particular species or landscape feature is more important than any other (looking at us *Homo sapiens*) as we all rely on each other in some way or another. It might seem small (a spider's web in the corner of your door) or it might be quite large (a hugely wide and free flowing river that supplies freshwater to a gigantic area of land and runs right to the sea), but they all play an irreplaceable role in how the world works as we know it. Each part is worth guarding, protecting, and allowing to be. As it is, as it should be. This anthology does not have answers, it doesn't contain the key to saving the Earth, but it does shed just a small but bright ray of light on the parts that make it so special to us. The small moments that remind us of how good it feels to be alive and stay alive in spite of the hardness, and our bonds to the Earth, to ourselves and the ones we love(d). Whatever that might be for you, it's what makes it worth the fight.

This is really what sustainability is all about. It is a collective movement, an ever-evolving opportunity to learn, to change, to evolve and adapt to the new world landscape we find ourselves in. Let the words between these pages bring you pieces of joy, recognition, remembrance, and consciousness for what you love most about being alive. Let them change you, lead you, and guide you towards a life that feels remarkable in its efforts to preserve what feels amazing to be a part of. Pass that feeling along to your friends, your family, to all members of the next generation, and let them carry that forward and forward until it's all we've ever known. Until the age of mass consumption and unbound expansion is but a small blurb in a history book, skimmed through and forgotten before lunch is over.

Keep fighting and take care of each other.

Ju Collins, guest editor
6/20/24

An Invitation

You've been here before, you're sure of it.
Over the river, through the woods, up and down the hills and down and up the valley slopes—the World swallows you and spits you back somewhere else, and here you are, standing in a meadow at the center of it all and wondering which path to take next. There are seven of them, eight if you count the way you came, and who knows if the trail is the same as it was when you stepped into the clearing.

You see the Sun, turning over in her lazy arc high over the crown of the leaves, but it tells you nothing of which route to take, which spoke on the Wheel to travel down next. Again you cast a look around the clearing, taking in the flowers, the flies, the mushrooms. You walk around the ring of trees, touching the bark of oaks and birches, trying to tell for certain.

Perhaps you can not be certain, the meadow looking as it does like so many of the others you've seen—but you are sure. You've been here before. Plopping down into the grass, you pull a handful of dried berries and a strip of mushroom jerky from your pack. Your teeth tear at it, seeking somewhere to put the frustration of the days of journey lost to another false start. You are about to wash the meager meal down with a pull from your waterskin when you look up, sure you have heard a knock. You look around but see no one, nothing beyond the occasional buzzing insect and the call of birds you can not see in the thicket of the wood.

Sure it was the heat of the sun, you are about to take another pull of water when you hear it again—a high, clear knock, as if of knuckles against strong wood. Fingers going to the dagger at your belt, you raise yourself into a cautious crouch, looking around and seeing nothing but the wind in the weeds. It comes once more as you stand to your full height, and you try to ascertain a clear source. After a long pause it comes again, and you are sure it's coming from the middle of the meadow.

Hand still on your sheath, you move across the clearing, the knocking coming in a slow but steady but rhythm, until, almost at the epicenter, it suddenly ceases. You wait for it to come again, but it does not. You look down, see only more reedy grass and nameless wildflowers, and kneel down, running your fingers through the green, against the dirt—dirt which, to your surprise, is only a shallow shroud. As you push away the brush you see it, the lichen-kissed face of a trap door buried below you. Hungry hands scrape away the foliage, clearing the earth from the ancient wood until the mossy planks are naked and exposed.

With a deep breath, you return the knock.

A yawn of silence, and then it comes, clear and steady as before.

The rusted ring is tethered to the wood by time, and you hesitate for only a moment before you use your blade to cut away the clutching roots. You slip your fingers around the metal and take a last deep breath of the muddy air. It opens far easier than you expected, thudding into the grass as you peer over the edge. Looking down, you see, impossibly, a sea of stars, the full Moon floating directly below. You can hear the buzz of insects just beyond, the trickle of water, yet more birds.

You observe the aperture, back up at the eight trailheads around you, back down at the hole in the middle of the World. Each mouth beckons in the trees, begging you to take their route away from here. You remember the instructions you were given—no matter the temptation, no matter how lost you feel you are, do not stray from the path.

And yet... you are tired of looping. You are tired of thinking that you've been here before.

Laying down against the grass, you look at the hole, at the yawning sky within, and reach inside to touch the night air. You see your fingers wiggle at the end of your arm, licked with the wind, and then a moth flutters into sight, and your hand goes still as it lights on your knuckles. A moment later it carries on, and you crawl forward on your belly, right up to the edge.

Without looking back at the woods, you take a leap of faith and dip your torso into the earth.

The air is as thick with fruit and flowers as if you have dived into a *pond, and for a moment aroma is the whole of your senses. As you grow accustomed, you realize that you are looking at two bare brown feet, toes flecked by clovers.*

"Need a hand?" *she asks, and you look up to see a freckled face framed by thickets of frizzy hair. Her skin catches the moonlight, throwing her soft cheeks and careful smile into a silvery glow. Her palm, rugged and worn, touches yours and you feel yourself pulled up and out, almost tumbling over as you land on the turf. You look down at the Sun far below, and then back up at the stranger.*

Smiling, she turns and begins to walk across the field.

You scramble to your feet and run after, drawing her glance as you catch up. Now, walking through the grasses and the flowers, you can see the lushness of the other side, of the wild Nysan terrain all around you. There are rolling rows of fruit trees, shrubberies and bushes and topiaries all around. You walk with her down no path at all, along the bank of a babbling brook, past mossy embankments, scattered patches of roses and rhododendrons and raspberry blossoms.

There is a basket over her arm, and you watch as she stops at a tree and plucks a plump red plum from a branch. It joins brown pears and fat grapes and peaches as white and round as the Moon overhead.

Maybe you know her from the stories, and perhaps she is a stranger, but before you can say anything, she turns to you. Her smile has stayed still since you first saw her, placid and firm in place. "I am that I am, if that's what you were going to ask." *She turns again, continuing her walk, but gestures to the tree.* "Go on, pick one."

So you do, pulling one down, the skin taught against bursting flesh, and as you continue your walk the juice dribbles down your chin.

"You may eat of anything here," *she says,* "taste whatever you wish, as long as you plant whatever is left—though perhaps do not overextend your curiosity. None of the specimens are labeled, not even the poisonous ones. I prefer to let them name themselves."

You search for words, find nothing that seems to be of any significance at all, and so merely take in more of the World around you. At the edges of the green you can see fields of wheat, of maize, of barley, waving at you from the horizon. Past the fields there are mountains, jutting up, wrapping themselves around this patch of paradise.

"You may explore as much as you like. It's yours, after all." She fixes you with eyes as deep and dark as the soil, and you realize with a sharp clarity that there's another smell, beyond the pollen and the pleasure: you can smell the loam below the grasses, below the roots, smell the wet, churning decay below that delivers life up towards the sky.

"Everything has a story here," she says. "Not just the plants or the fungus—the insects and arachnids, the stones and the sticks and the things that used to be alive and the things that never were and so always are and always will be." Her smile returns, and you wonder what mysteries that it, like the soil's mask of greenery, might hide.

"Go on," she says. "Seek what there is to seek, and look, and listen. Go on out and play."

What I Ask of the Future

by Tricia Knoll

Scientists predict auroras over Arizona as the sun crackles record flares. I anticipate more work burying whales. Red stars on barns are outed as fake news. Pollution causes white-leaf clover to sprout a fourth leaf; cynics rhyme words with luck. Some want to seed the Arctic with small silicate-glass spheres to reflect sun back into space. The forests? Scientists cinch ultrasound belts around trees in Munich to determine which species can handle drought. A word used as if it can conquer chaos is resilience. Solastalgia appears in crossword puzzles.

>   I think that I shall never see
>   a creature as friendly as a manatee
>   I never will see.

This fine April morning I listen to the Dalai Lama chanting an ancient mantra for defeating death and increasing prosperity. I meander a slope overlooking the big lake and rolling mountains. Canada Geese pick up the soundtrack. Some call future a river we admire from the sideline, shilly-shally in the shallows, or dive into headfirst. We wear blindfolds at the window that partially reflects the self.

I find a robin on this spring-green lawn, stick legs in the air – it collided with that window.
A too-early-out honeybee hovers. I cannot forget last frozen winter, the repetitious wheel, five times in a decade a100-year snowstorm.

Future is my topography.
Steadfast solidity of cliffs.
Rubble of weathering and erosion,
discontinuities and slip faults.

I find a kitschy polished white stone with *hope* carved on one side. To pass hand to hand. Or take home buried in my pocket. Uncertain of the sentiency of rock, I place it in the bowl-lap of a cement frog meditating, hands on his knees at the doorway to the sanctuary where I hear this chanting: *so, I am not separated.*

Last week three poets read their poems about doomsday from climate change. Each mentioned grandchildren and offered nothing to put in their pockets. Never mentioned great-grandchildren.

Geology. Long-time topography. Oceans rise uninvited. Boulders breathe relief from burdens of gone glaciers. Wind wipes away the built world. Rivers go wild with snow melt. Valleys recover from war. Tree species migrate north. Om and abide.

Two Poems

by Kostandi

Nature Lover

I am in awe of nature
But revolted by myself

I marvel at mountains
Yet frown at my body's hills and valleys

I love lightning striking across a blackened sky
Even though it mirrors the varicose veins on my skin

In the ridges of my clammy thumb
I see the rings of a mighty oak

My cracked lips that I gnaw at nervously
Remind me of the shedding scales of a snake

Even the perspiration dripping down my brow
Tastes the same as ocean spray

A tiger's stripes, sweeping across its fur,

Are reflected in the marks stretching over my thigh

I know my aging cells will one day
Explode into a supernova

How can I love nature
And not include myself?

In The Afternoon

The afternoon is when
The clouds tumble out
They billow, white like cotton
Somehow even brighter than your smile
I imagine touching them
How plush and peaceful they must feel
But they're always out of reach
Just like you
The afternoon is when
The wind dances
It whispers past my ways
Twirling the air into a gentle, constant current
I purse my lips
A weak attempt to replicate the sound
Because I still can't whistle
Not like you
The afternoon is when
The water becomes darkest
It's navy color borders on black
Rich indigo surface, waveless and mysterious
I muse at how deep it must go
Cloaking creatures and treasures
Hiding secrets I want but will never know
Same as yours
The afternoon is when
My loneliness drags me down
Like cement blocks clung to my feet
Unable to wrench myself from the grit of the beach
I try to move forward
To shake off the sand and salty air
To free myself of this heavy mass
Like the memory of you
I carry the weight of you
In my hollow chest
Coal to my fire, we burn

A speck of red against
The water's ultramarine
I can only burn so long
I want to, try to release you
Cast you away
A skipping stone across the ocean
But you stay in my chest

Before the Storm

by Paul Hilding

    The dogs and I don't go as far as we used to. Instead of the half mile jog to the mailboxes at the top of the road, or sometimes to the river another half mile beyond, it is all Cabo can do to limp alongside me for a few hundred yards. Loma stays close in the alfalfa field, chasing geese and checking scents left by deer and other late-night visitors. But he's started to slow down as well, his hips stiff with arthritis.

    Our unhurried pace allows me to study the weeds along the roadside. As fall becomes winter, I watch the goldenrod fade to cream, and then grey. The same cold weather alchemy transforms the grasses and mulleins to muted shades of hazel, rust and beige. Scattered rows of curly docks have turned entirely brown, as if cast in bronze from stems to leaf tips.

    It is easy to succumb to melancholy during this season. Surrounded by so much death, it is natural to think of the short time we have left with our aging dogs. And, after my diagnosis, with each other.

    But there is also beauty in this stark landscape that draws me out each day into the December chill. And something more. A meaning, perhaps, to be worked out.

Now, as dusk approaches, I pause at a desiccated stand of wild sunflowers. Their straw-colored skeletons remind me of candelabras, dried blossom heads tipped upward toward the slate grey sky. Storm clouds float low in the western sky like silent battleships, crests tinged blood-orange by the last light of the winter sun.

Cabo takes advantage of the break to sit heavily at my feet, seeming to admire the scene as much as I do. After a long moment, he slides to the ground, sighing in a way that seems almost human.

I kneel and stroke his head. His sad eyes meet mine and, not for the first time, I imagine he understands it all, my own future as well as his. My symptoms had returned a few days ago and the meds were not working as well. "It's a progressive condition," the doctor had offered by way of explanation.

The horizon darkens. The battleships sail closer.

Eventually, Loma sniffs the approaching storm and looks over from the stubble-strewn field. The lifeless stalks shiver in the rising wind. I whistle softly and pull my jacket tight against the cold. For today, it seems, this is the right place to turn around. The dogs wag their tails as we start the short walk back to the warmth of the cabin. And I feel grateful as well. For these moments before the storm. For all of these moments, every day.

Before all of the storms.

Two Poems

by Claire Thyne

proliferation

why bother to cut the grass? and,
no offence, but fuck you for
drowning all the dandelions.
sacred ground cringes at
the touch of your volition.
you are more a weed than
creeping woodsorrel, claiming
undeserved territory under the
rouse of a civilizing mission.
and just when you think you have
annihilated this disease,
watch it bloom like bombs.

THE BEST MANIPULATORS ARE THOSE WHO MAKE BEING LESS LIKE YOURSELF FEEL LIKE PROGRESS

Meeting the Forest Floor

I have never seen eyes burn as brightly

As when I swiped the head clean off with a

Samurai sword. I watched you wriggle and

Your eyes burned brightly and you looked so pretty

Lying there. Dead in the middle

Of the forest. A thrush brought it's little

Prey over, a snail, and knocked it against

Your forehead once you fell. Your body wriggled

But your head was perfectly still and so

Your skull helped the thrush crack its little prey's

Shell open. When the snail was sucked up, the

Thrush flew away, and your eyes still burned bright

Like a lighthouse beacon. I would have liked

To see the light leave your eyes but I grew

Tired of waiting. Even now I'd wager

A pretty penny that they're still as hot

As the day they saw the trees touch the sky

For the last time.

Brassy

by Keiraj M. Gillis

A rogue hollyhock grows in my yard,
blushing each spring with downward-facing
bell-blooms in a sloppy phalanx; we have
an agreement—the plant and I—that, yearly,
I'm permitted to pluck no more than one
of its rosy bells for my own. And in honoring
tradition, I reached to shear a shy little one.

A honey bee, stout and bossy, objected.

It zipped toward my arm, parachuting onto my
wool sleeve, attacking but failing to pierce
the cloth. And I thought, *How audacious,*
that a creature would feel emboldened
in its indignation to attack a thing a thousand
times its size. But disparity in size was no
deterrent for it, apparently, when fighting
a battle it determined *must* be fought and won.

Such is the rage, I suspect, of every tiny thing,
as it is ignored, displaced, and stolen from.
Only when brassy—confident of purpose with
dutiful bombast—is their roaring ever loud enough.

Four Poems

by Diane Funston

Taken in Small Measure

Trying them for the first time
kumquats at a tea room
in a salad
A miniature orange
I thought
Biting in to the whole
Feeling the tang
the almost-sour
surprise on my palette
followed by sweetness

Today I have a tree
in my front yard
under the window
Sweet blossom scent
wafts inwards
Biting into the harvest
the reality of both sensations
sweet and sour

taken in small measure
moments in a lifetime

Bittersweet

Climate Changes

Deep in El Junque rainforest
on the island of Puerto Rico
we rocked in the hammock
listening to the tree frogs
Coqui, coqui,coqui
a lullaby in the mist
where we had time
to watch the ice in our drinks
shift in the melt from the warmth

In the August winter chill
the wineries welcomed us
in the fertile Hunter Valley
We bought warmer coats
and sat in the upper outside seats
of the tour bus in the alienated city
just to catch the street lighting at night
see the old churches by day

After two more years in survival mode
the fire that took us away from a garden
to a suburb of rigid rules
then return to a home rebuilt from the ashes
a trip to help out in a very dry New York State
we've been grounded in our travel
but not our dreams

To sit and remember how it was
before the closings and quarantine
all the un-natural disasters
Before—
when we flew away like lovebirds
tasting the nectar of travel
and the privacy of one another
without interruption or rationing

I now sit alone with my gin and tonic
from the gift of our own Meyer lemons
I think of new places to see
when the masks come off
I make a list, beginning with the tropics
I always look for heat when I write
if I can still find it

The Secrets They Told Me

Nothing is empty
not the rain-cleared skies
the growing winter garden
Every step is full possibility
we deep dive
into our pasts
but that's the most empty
of who we are presently
I've buried the dead
and shoveled those alive
down
far deeper

I've tread sidewalk
and cobblestone
walked while weeping
and singing loudly
the mercado in Zamora
where street food
beat the linen tablecloths
of the nearest big city
I've known all sadness
until this year
this solid calendar
of celebrations

I've mourned no one
close this year
Some passing at ages
that inspire awe, yes
those living large
their best lives
those who truly
contained. multitudes
of things bigger than

themselves
bigger perhaps
than all of us
who still breathe

That's the secret
the mushrooms
and Mexican parrots
told me
in deep forests
and bus rides
where tourists can't go
even at Casa Azul
where Frida made room
for Diego
before he ate
her barbed wire heart

On this day

a river riots through my yard
The once-dry stream rushes
with seasonal wrath
a water dug trench
deeply carved cleft
spends a life force
trying to cheat
this high desert

I sip herbal tea
as serpent surges
with sound of fury
floods my land-locked
reverence for all
things in transition
I am reminded then
that even eons of rock
are changed by water

False Spring

by J. J. Stewart

 The backyard looked like Hell on this 3rd false spring in the middle of January. The rains had been and gone and been again, leaving my small patch of California Paradise seriously rethinking catholicism with its Old Testament roots. This year's greedy avalanche of crab grass was already entrenched, warring with sour, daffodil-headed oxalis and smothering the poor peasant stock,hard- scrabbling California-native lupines and pennyroyals, and delicate five-spot underfoot. We'd ignored… Everything… For far too long, preferring indoor lives full of holiday over-indulgences: wines and soups and binging as many K-dramas as can fit between an early sunset and a late bedtime.
Midges and mosquitoes sang through the air, warring with hundreds of silken threads of new hatched spiders, none of whom would live reasonable lives when Winter reclaimed the neighborhood. Was I really envious of a mayfly's 24 hours where this single spring day, sandwiched in the midst of dreary skies and clinging mud was all they'd ever know? Woken from their eggs too early, to live too quickly, to return to unknowing without understanding the inevitable consequences of tomorrow?

The cats wriggled through my ankles as I stepped out into the open. They were too lazy to escape and too inept to hunt, so they were allowed the privilege of their curiosity.

The birds certainly couldn't have cared less, chittering and chirping, flitting from the armored safety of the 80-year old yucca to the brown bundle of sticks that I fondly imagined was a thriving cherry tree, to every greening bush and shrub, scratching in the dirt, gorging on half-sprouted, water logged seeds that I'd remembered to put out for them, weeks ago. Before the rains came. The birds and I put our faces to the sun as much as possible as we all knew what would come tomorrow. (Climate change.) The towhees with their dramatic black heads harvesting abandoned sunflower heads, nuthatches calling, kicking away millet for better-prized poppy seeds. And cavorting from the roof to the neighbor's oak to the empty lilac, a pair of scrub jays, screaming for peanuts. A territorial hummingbird, peeping its ego, hovered in my eye, staring me down, begrudgingly accepting my oafishness as I stumbled through my molasses existence.

I picked up my metal trowel and my plastic bucket to wage a bit of tepid war against the weeds. To shove some electricity into my muscles and dirt beneath my fingernails. My bucket filled quickly and the earthworms gave me the finger from the deeply disturbed earth before wriggling down to rebuild their homes. Penwiper, tuxedo tail bristling came to inspect my effect.

"Mrrrl," he opined.

"It *is* a very little result, " I admitted. A bucket full of weeds equating a single foot's circle of clear soil. "But I am old and tired and this is only a hobby."

My cat, unimpressed, slitted his bright eyes against the brighter sun and picked a fight with a white cabbage moth.

And lost.

I picked him up and brushed the moth dust from his nose as he lashed my ribs with the whole of his indignity.

He leapt down to tell tales of woe to his reluctant partner, who couldn't have cared less for him or his complaints. She turned back to her own obsession; gnawing on as much grass as possible. To be vomited up, at a later date, on our rugs or hardwood or freshly made bedspread to be found right before bedtime.

Penwiper, with no angle to sympathy, sloped off to the picnic table to console himself with scraps of sunlight and delusions of righteousness.

I turned to Arjumand to say "If history is written by the victors, I suppose anything can happen if you never learn to read." Arjumand looked at me and horked up a wad of semi-chewed verbena on my shoe.

And so much for philosophy.

The sunlight wheeled overhead, lengthening shadows to timepieces. The tomatoes, which had been wearily soldiering on since summer, their vines still full of green and blushing fruit, shivered and moaned in decrepit whispers, crying out for their well deserved rest in the Eternal Compost Heap. Tomorrow, I lied. I knew that, by then, I'd have been too tired, too stressed, too much a societal survivor to step outside and attend to my responsibilities. (It is, after all, only a hobby.)

I heard the grackling of Ravens overhead and, peeping my face to the sky, saw them, high above, cavorting free and pulling storm clouds behind them as they gamboled towards bedtime.
The lesser, bright-colored wild birds had fled some time ago. The midges and other insects fell with the temperature as the light faded. Strands of spider silk turned to invisible traps for unwary prey and blundering humans alike. The cats whined about autonomy as they climbed the old, moss-speckled brick steps towards their heated blankets and nutritionally balanced dinners.

"Life is just like that, sometimes," I told them as I shut the red wooden door on the chores and outside agendas of the natural world and turned towards a very full evening of munching food that I'm almost sure I'd sworn off on New Year's Eve. If only it hadn't been a Day with Bosses and Questions and the need for sanity-surviving treats.

And we all tucked away, the cats, myself, the Ravens and the midges, the sparrows and the tomatoes, all of us on this 3rd False Spring Day of January, down into whatever passed for comfort as we waited for the storms to resume and for Winter to rush back like an ex-lover resuming an argument that you'd thought long settled, that we all had to tremble through.

Three Poems

by William Doreski

Living Rough

While tornadoes ravage the south
and freezing rain licks the north
I travel west to learn to live rough.

No clean clothes, razor, or books.
Nothing between me and the world
except flatulent yellow breezes.

Every month is the cruelest month,
sapping me with hugs and kisses
I would never dare reciprocate.

I sleep in oil drums by the railroad
or in discarded packing crates
behind industries closed forever

on the outskirts of Cleveland,
Akron, St. Louis, Carson City.
I hide from weather in the mouths

of abandoned mines, under cars
junked and rusting in the desert
where nuclear fission blossomed.

I gnaw on creosote lumber.
I eat weeds and challenge cattle
for the contents of their feed troughs.

I drink the most polluted water
in hope of destroying my brain
to shuck the dread that drove me

from my house in New Hampshire
where the snow heaps like winnings
at gangster-owned casinos.

The days delaminate in colors
properly housed people never see.
Mauve canyons fill with sundown.

I've followed the railroad from there
to here, the steel honing itself
to cut a million gasping throats.

The howl of straining diesels
inflames my remaining senses
and I begin to think again.

I peel my shadow off myself
and send it to the nearest town
and order it to survive me,

marry well, start a small business,
and redeem itself from the dark
I've carried so long in my pockets.

A Bogus Portrait of Shakespeare

From my fifth-floor rooms the view
across parking lots to gaudy
night-lit glass office towers
suggested that nature had died
of old age, surrounded by loved ones.
Sirens unzipped the reddish dark
embellished with spatters of blood.

Every day in sunlight or snow
trudging to work in a cheap suit
 I slipped through the gleaming streets
with my defenses down in hopes
of inhaling harbor salt-smells,
blowing a mile from the east
to add to my stock of horizons.

Fifty years later I'm perched
in a rickety wood, my house
fragile as a pressboard cigar box.
Nature staggers around with sap
leaking from a million wounds,
negating my lifetime of work
by insisting on gray sentiments

like those Keats had to overcome
to write about urns and nightingales.
I picture him on the Isle of Wight,
a bogus portrait of Shakespeare
tacked on the wall of his rental.
I've never lacked grave distractions
but haven't wrought such gnarly words.

The woods rattle and threaten me
with dropped limbs, toppling trees
while the ghost of my glen plaid

office employee still waddles
to work all day at a metal desk,
an unread newspaper sulking
in the lower right-hand drawer.

Two Cherubs

Two cherubs painted on plaster
in a ruined summer house.
This outdated motif critiques
the close of a feckless year
so criminal that history texts
still unwritten are crumpling
their pages in verbal dismay
at war and corruption expanding
to include the oldest pensioner
and the newborn gladly nursing.

This summer house once belonged
to the estate of the man who devised
Clue and Monopoly, board games
that bored me right through puberty.
In deep snow there's no evidence
of the formal garden that thrived
through a hundred winters to lilt
back to life each April, grinning
with crabapple blossoms and vines
as thick and tough as pythons.

No evidence but the wrecked
summer house where I shelter
from the northwest wind fumbling
for victims ripe for frostbite.
The cherubs bear between them
a banner no longer legible.
But their faces retain big smiles
against a background of pastel cloud.
Those smiles could sell anything.
The odor of seduction lingers

from drunken parties years ago
when the big house presided
over New England's simple wealth.
Now converted to condos
the mansion cowers as if caught
naked with the hired help.
I wish I could take these cherubs home
but the plaster would crumble if touched,
leaving me undefended against
even the slightest warp in time.

Orchard Language

by Kathy Pon

I.
Orchard language is viscous,
thick with bloodline, sticking to
the tip of a longing tongue. It
bonds to legacy, the cultivation of
soil, seedlings pleating acres
staked with white cartons.

Before we marry, I meet
your mother, Machi. I realize
her first language is not Basque,
but farming. I hear in her voice
a steady cadence of the ranch
that nurtured you, fruit trees invoking
your family to raise up agriculture
like a host to their lips.

We mortgage our safe suburban
beginnings for dirt of our own,
plant an almond orchard.
Immerse ourselves in

the linguistics of growers.

II.
Orchard language is lush
with leafy unfurling
of green, afternoon's spray
of light stretched long
on the treeline. It speaks in
lilting branches, birdsong cascading
from canopy foliage, dragonflies
droning daylight.

By listening I absorb vocabulary
of farming. In raindrops grazing
leaves, soaking deep into terrain,
I understand *aquifer*.
In the silent snore of dormancy
building towards bloom,
I comprehend *chill hours*.
I recognize groans of fullness,
almonds splitting leathery hulls
as *harvest*. Pronouncing time
to *shake*, I marvel in the soundscape
of drumming nuts falling to earth.

To orchard is the conjugation of breath
giving life to our almond seedlings.
Cultivate the soil, command
the sun and water to transform
energy into nuts. Nurture them
like our own children.

Three True Stories

by Cassandra Caverhill

WHILE BIKING WITH MY FATHER WHO HAS ALZHEIMER'S

I didn't know how far I could pedal when I've struggled with self-propelled momentum as of late. Dad let me take the sidewalk ahead of him, moreso because he needs me to lead now, and I felt like I was ten again: wearing my helmet, checking behind my shoulder every few blocks to make sure he was still there, watching over me. We took the road until it ended, a new subdivision of modern homes in black and stone rising from former farms that existed when we were both young.

We stopped at the beginning of the bike path. I sipped from the bottle of lukewarm water in my basket, offered some to him. He removed his ballcap and swiped at his wet forehead, waving me off.

"Should we keep going?" he asked.

I didn't want to turn back just yet, but I didn't want to push him too hard in his condition either. I noticed a narrow dirt trail wide enough for wheels on our left and wondered where it led.

"We can take it," he offered, still retaining direction, true north. "It's a shortcut to the park."

He pushed ahead of me though reeds, horsetails, and scouring rushes—soft walls granting us passage. Dainty, white anemone and pillars of arrowgrass swayed in breezes that cooled our sticky skin; butter-coloured artichoke and purple-tinted bog aster dotted the hillside.

I swallowed the sweetness of summer's light, the sky overhead profuse with soft heavens. The wildflowers bowed in reverence toward my father, as I watched him drift ahead, unaware of the tears that soaked my cheeks. The victrolas of foxglove and bellflower playing a static and hissing hush of love.

DIRECTION

The sky was spitting in my face. I entered a trail and saw a ribbon of coloured coats ahead, paused at a clearing, pointing. I decided on a different direction to be alone with my thoughts. Along the fenced division line, sparrows cried in the crowns of ancient oaks. I stopped and cried too. Woodchips gave way to sand. Overcast chill to scorching sun. And in my mind, he and I were on the beach once more, listening to the powerful lake give and take. I opened my eyes, and an unleashed mutt watched me, its owner calling out to "wait." I wiped my face and carried on, grateful that my wooly muffler covered me from getting too real with someone else. Trains shunted outside of the rice factory, clangs carried on wind, and transport trucks whizzed by the protected woodland. I smelled ash and saw the residue of soot and smoke. The fields were charred, with tendrils of salt remaining. I know sometimes that one must burn down what's living to move beyond survival and into thriving. I just didn't expect to feel this way, doubling back at the precipice of change, fear rendering me just as weak as I've ever been. I crossed a footbridge covering a creek of rainwater. In the thicket of my heart, a doe emerged and studied me with an unshakable stare. I waved a mitten. She shook her tail. Then, a second doe crossed into a clearing, pausing out of mutual acknowledgement before she passed. And finally, a third came closest yet, nibbling on young grasses, unafraid. The three of them gathered inside a field of clattering trees that had not yet budded. A crimson flame pulled me from the wood brush and fallen leaves: A cardinal perched in a young birch. I noticed a wooden stake in the ground before it, carved with a continuous arrow pointing in the direction I had come from and the direction I was heading.

IT'S JUNE

A landscaper blows a loud tube of artificial wind at baby blades, ushering green toward gutter. Dropped samaras stare up from the sidewalk like the eyes of Horus, protecting hidden growth on these shady streets. Earlier, inside, I studied a housefly at rest against the window: Did it know that surface as permeable? Or was the mere image of outside enough to suffice its desires? An exterminator consecrates the corners of a clapboard house, spritzing from a metal jetpack to kill whatever festers in the dark. My horoscope says it's a good time to act on impulse and so, these steps are the beginning of letting go of the life I've known as his wife. A red van hurries to a stop sign with the intention of rolling, that is, until the driver notices me moving through crosswalk. And when he eagerly waves me on, I mutter, "Don't fucking rush me."

Another Revolution

by Robert Witmer

The fetish wore two masks—tragedy in front and comedy behind. We put him on a potter's wheel and dreamt of civilization, an Earth Sounds LP on the turntable.

Seven Poems

by Svetlana Litvinchuk

Just Say the Word

I survey my garden, seeking evidence of progress, searching
for some measurable growth in the face of searing sun. I calculate
the diameter of rose pollen. I count bumblebees napping inside

the arms of flowers. Tie pea plants to trellis, encouraging them
to skyscraper. Mound potatoes into monuments. In the torrential rains
of spring, the pelting hail, the tornado winds ravaging my dreams,

space opens up amidst the failures of my cultivated future,
the too-soon plans that die before ever bearing fruit. I am surrounded
by the second chances I don't know how to ask for. Will the Earth

be this forgiving after seas boil into fish soup? After my humid
subtropical climate steams itself into desert? Collectively, we lurch
toward the cliff of civilization. It is the only thing we do in unison.

Near the edge our dancing feet crumble mountains into pebbles, threaten the whole backbone of the earth into collapse. So, I keep planting seeds, throwing clouds of would-be flowers into parking

lots, hoping to redeem my glyphosate-spraying neighbors. How many good deeds are enough to keep the scales from tipping us all into the mouth of some terrible God for maybe just long enough

to keep this Earth an Eden for the duration of my daughter's lifetime?
And maybe her daughters? Okay, and maybe also her daughter's daughter's? Tell me how to live long enough to forge some kind

of relevance. To make some historic suffering worthwhile and hold our missing pieces together into wholeness again. How do we live with ourselves knowing we leave to our children less than we've

ever had? I'll sacrifice everything, just say the word. I'll walk through tidal waves to make my daughter's journey a safe one. I'll live on sand and dandelions to keep her belly full, drink up the mud

of rivers to keep milk flowing into her. I would trade all of life's wonderous surprises to give her just a few predictable happenings to count on, to bequeath to her a planet that promises not to swallow

her, that is willing to accept my apology on behalf of my brothers and sisters to forge some unbreakable truce, please say it's not too late.
Just say the word.

Red Planet

Go read poems about living and find that we all agree
our Earth is dying. What repeats is red. What repeats
is rage. Go scan the news and find an image of a suffering

ocean, the one from which all life sprouted legs and walked
out not looking back. No one calls this red water blood.
We drown it in plastic. Make soup. Drink the grief of extinction.

You knew about Jung. About our collective dreaming in red.
It is spring and red appears inside our closed eyelids when
we should be writing of green leaves and crystal meltwaters,

of glaciers calving and reproducing to suckle and quench
our progeny, of flowers blossoming in the hearts of maidens.
Instead, our throats open and out pours sand. Instead, we write

of poisoned fish in the garden. Of island nations swallowed
by a raging sea rising, bubbling with heat, lapping at us
with sharp tongues that admonish our father's choices, gradually

rising to pull us under like the next generation of Atlanteans.
We write of the ghosts of burning forests. Of bears relocating
to suburbs as their natural habitat retreats northward like honey-

loving icicles who simply exist for love of living, knowing
nothing else but this life. Reminding us that to continue living
is all life knows of itself.

Earth's lungs emanate smoke, inhale it, suffocate in red
flames, as if the Earth has always been burning on the inside,
smoldering deep in her molten heart since the day she was born.

Flames- man's greatest discovery- lick our sleeping feet.
We suffer a collective fever flooding our hot metal dreams,

warnings of our shared fate. For that is mankind's real greatest

discovery, one we forget again and again: how truly connected
we are. The red tide is swelling over into our dreams now.
It pools between our ears, makes a harbor for dying stars,

unresponsive to the passions of a moon. Our pineal glands drown
from lack of connection to each other. The Earth gives.
We extract extricate excavate roots as the trees try to show us:

Look, we hold hands under here. Life is a continuum, a carpet,
we weave it as we go, we must hold hands to stay afloat. To float
along the universe we construct a raft. To let go is to unravel

is to collapse into strings is to drown in the vastness of life
yearning for itself. We have always gazed upon a shared moon.
The only one we have from the only Earth we have. The abyss

stares back into itself. How cold it looks up there, cool and crisp
and free from that white gauzy cloud-blanket and now when the moon
gazes back at us we see it painted red. We pray for Mars,

for our salvation. We dream together; we must dream together.
We mistake satellites and space junk for shooting stars. Our wishes
heard but unanswered. Our prayers unanswerable.

Sorry for the Interruption

I'm sorry for all those years of
unbridled enthusiasm
for my knocking on thousands of doors
(I had a signature knock)
so pleased to be interrupting life

Those *No Soliciting* signs I smugly ignored
(I'm not selling anything,
just doing my part to save the planet.)
Forget your dinner getting cold
(are you aware the ice caps are melting?)
Forget the argument with your wife—
(there's conflict in congress, we must
write to them NOW.)
forget your disabled grandmother unable
to rise to answer the door
(stand with us, there's always power in numbers.)

And all those other signs I disregarded:
(Baby sleeping, please don't ruin this for us)
but you can just get her back to sleep
(can't you just—)
forget your bills and write us a check
one for every month
if you really mean it when you say
you care you really do.

I thought it was so simple
I didn't understand.
Idealism is like that:
It doesn't understand when
things aren't that simple.

But, why aren't things simple?

I mean

things should be simple.

Tornado Season

seems to signal
a time of furious growth
the change of unfurling leaves
the grass stabbing through
soil's softness
sprouting seeds clawing apart the ground
to be healed over by the lushness of
flowers in our cracked hands

summer blows in slowly and then
all at once until the air stills
as if it's always been this way
the constancy of cicada trill that
reminds me too much
that everything knows when it's time
to resurface
that nothing worries about the part
about staying

Let What May

Let what lays at my feet
grow toward the light
and become tomorrow's flowers.

Let each seed resting
in my right hand become
a tree-lined path marking
all the places my feet have touched.

Let my passing shadow
shield some small thing
from the noon sun's scorch.

Let some small miracle
be sung from my lips
into the ear of some small god.

Let what may come for me, come
finding me on my best day.

Let beauty happen to me.
Let beauty happen because
I was here.

I Am Not Separate

I am the breath that moves through me
not separate
I am the tidal wave
the joy, the grief arising from the same well
I am the bellybutton of my own universe
as both the breeze and its destination
I am the three words strung together that I call my name
a combination
I am the strand that connects my ancestors to the future
not separate
I am a root seeking nourishment
I am the soil that gives
the pieces that turn to crumbs to feed the trees
as they make, among many other things, oxygen
I am the breath

Grinding Down a Tree

Today I saw a tree with all its limbs scattered on the pavement, last year's dried brown leaves still clinging to its branches, clinging to some hope of life. Three men were grinding out its stump and I felt the pain deep inside the churn of my stomach down where a seed still remains outside my womb where the love of all living things refuses to die no matter how much pain there remains in the world, no matter how much cruelty we inflict on one another.

It was healthy, wide and tall, still in its prime, had done nothing wrong as far as I could tell. Frozen with grief, I watched this perfect tree being murdered- we all did- there were so many people around and it was broad daylight and it was absurd because no one shouted or did anything to stop it but I screamed on the inside it rose into my heart and inflated it like a balloon filled with all the Earth's music-tuned-elegy and I wonder now if it was the tree's voice echoing in me. I think maybe it wanted me to speak for it, was asking me to say all the things that it still needs us to know, that we fail to learn like the fact that even a dead tree can be a home.

In its dying cello groan it reached out to me despite limblessness and in the way that we all do it grabbed deep down in me where some undissolved stitches somehow remained from some long-ago scar and it found the edges and began tugging and tugging. The balloon of my heart exploded with a pop and every single star in the universe came tumbling out and the pavement was wet and painted black as an eclipse as I stood there and I think maybe it was looking for a new home in receptive soil, was simply making space. So, I said come on in it's warm in here, but please be careful, I have roots of my own.

If the three men encircled it while holding hands they wouldn't have been able to hug all the way around it; it was too much for them to love. But not too much for them to kill. It nearly burned the machine's motor out, you could smell it. I wish it had. I heard the machine groaning as if it didn't enjoy the work, as if it wasn't made for this. It nearly dragged my roots out and chipped my heart down until it was the soil and the seed in my stomach planted itself in its place amid the stars.

Poem on a Hummingbird Sage

by John Winfield Hoppin

Ah, fertility!
The garden blooms
Purple-throated hummingbird sweeps down
for a sip of the purple hummingbird sage
is startled by me
zips off

Nature machine rolls on
another round
another chasm
it's beautiful

Ah, Sunbeam!
Great to grow a thing perfectly,
to make a thing be perfect

Purple-throated hummingbird and
blossoming hummingbird sage
and sunbeam
Things that've always been around
but never before in just this way and
occurring before me now
A carpet of blessed reality rolls out into infinity
& this is happening everywhere
any place you go

A high five
to everything
in this dimension

JESUS COMING SOON
An Essay

by Katie Bausler

 To the creeks near our home, salmon return to the place they hatched from symmetrical eggs. After four years of surviving predators and dodging fishermen, some appear out of energy to swim upstream. Their stinky carcasses litter the creek and bank. But new life is underway. My dad leans his gangly self over the side of the bridge, fascinated as a four-year-old by the fish pairing up below. "That's the female wriggling a bed in the gravel for her eggs," I tell him. "The one shimmying at her side is the male who will fertilize them." Going on 93, dad's largely lost his short-term memory. But he is all in this moment, witnessing chum salmon, coming home.
 In the coves out the road, you can hear the guttural breaths of humpback whales rising to the surface for a hit of oxygen. These same whales spend the winter fasting and mating in the waters where my brother takes his daily swim, off the Hawaiian island of Maui.
 Three years ago, my brother and sister-in-law sold their home overlooking a freeway. They left San Diego for Maui, a plot of dirt and a wooden shell that would become a modest townhouse in a

sprawling housing complex in Lahaina, a place historical and sacred to its original inhabitants.

In mid-July 2023, dad and I flew across the Pacific Ocean to visit. My brother and his wife seemed more relaxed than ever, easy smiles on sun infused faces. The first place they took us was an airy restaurant plopped on the edge of the ocean. The back "wall" nothing but white muslin fabric, filtering out the late day sun and letting through silhouettes of sailboats and paddleboards. We savored sea salted mahi mahi and sweet mango.

The bistro is just north of what the locals call Baby Beach, where Native Hawaiian families pull their trucks up to the shore and set up an outdoor kitchen on the sand. The inviting scent of fresh grilled fish wafts in the warm air, while three generations play in the shallow water. A short stroll away, a teenager practices his swing on a makeshift mini-golf course in his grandfather's beach backyard. Next door, couples tango at an open-air dance class as the sun sinks into the sea, turning the entire scene golden. A lone surfer rides a series of waves just high enough for a long, laid-back ride to shore.

This must be paradise.

Paradise is the name of a town in the foothills of California's Sierra Nevada mountains, population, 27,000, decimated by the Camp Fire in 2018. Eighty-five members of that community lost their lives.

I am the first of six siblings who grew up in California, not far from the 2017 Wine Country Fire that wrought searing images. The sun merging with the edge of a burning earth. A little girl with stringy hair and a soft face, clutching her charred bicycle where her home once stood, grey ashes and black soot, the moment her life shifted. A Santa Rosa neighborhood as if it'd been bombed, save for one intact cul-de-sac, the roofs clean and the lawns green.

In 1971 my parents bought a house in a cul-de-sac thirty-five miles south of Santa Rosa in a place called Terra Linda, beautiful land in Spanish. My high school alma mater was an evacuation center from the Wine County fire, victims taken to the hospital down the street.

My husband and I raised our two kids 2500 miles north along spruce, hemlock, and cedar trees in the largest contiguous coastal rainforest in the world. Here it rains so much, the weather service is challenged for descriptions. Scattered showers, widespread showers, rain showers likely, and the ever ambiguous mostly cloudy with a chance of rain. But to us, the mountains, glaciers, fjords, and islands inhabited by bear, salmon and mountain goats feel like paradise.

Lahaina is set on a narrow strip of land between a desert-like mountain canyon and the Pacific Ocean. Every year more than two million tourists visit this compact municipality of 13,000. Bumper to bumper trucks and SUVs snake along the main drag, Front Street, while tourists crowd some semblance of sidewalk. On our week-long visit Dad and I snorkeled with bright young fish and swam with a sage sea turtle.

Enroute to the Maui airport for our flight back to Alaska, I asked my brother to pull over. I hopped out, compelled to photograph what caught my gaze each time we passed it; the confounding words rising on stilts, all capital letters atop the church where Front Street met the highway: JESUS COMING SOON.

The sign inspired a track on the societally self-aware album

Hotel California, released by The Eagles in 1977. On "The Last Resort", Don Henley croons about escaping the shallowness of Los Angeles for Lahaina, where the missionaries left the neon sign.

We returned home to a narrow strip of land nestled between the saltwater Inside Passage and the Juneau Icefield. While about 30,000 people live here, more than one and half million tourists disembark from cruise ships from mid-April to mid-October. They ply our main drag, South Franklin Street, built when gold mining birthed the capital of Alaska.

The following weekend, melting snow draining from the receding glacier north of our house overflowed a lake and re-routed the river downstream. Residents watched in awe and horror as the water temporarily rose, eroded banks, and took out old growth trees, one at a heartbreaking time. A video that went viral shows a house collapsing all-at-once into the river. Homes of our friends and fellow community members were flooded, condemned, or deemed uninhabitable.

The following Tuesday, my cell phone buzzed with a call from my brother. He sounded scared and shaken. "We're almost out of gas. We just drove through palm trees and power lines on fire, falling around us," he relayed, between breaths. His wife chimed in. "Katie, I've never seen anything like it, the black, billowing smoke." They were pulling into a parking spot at the next tourist attraction north.

They'd barely made it out of Lahaina.

My brother and sister-in-law were fortunate to escape a literal hell on earth, where at least 100 people perished, some in their homes, or in cars stuck on the bridge I'd cruised along on a bicycle just days before.

A highlight of our stay was the celebration of my brother's birthday at a lauded Front Street institution: The Lahaina Grill. A bright mural on the back wall reminiscent of Monet. Our attentive waiter Paul Newman's double. Six of us gathered around a long table with white tablecloths, toasting the birthday boy with sparkling stemware. The table came with a roving staff photographer who snapped a lasting memory of our delighted faces.

The restaurant has disappeared now, along with the rest of downtown Lahaina, wiped out by a cauldron fueled by hurricane driven winds blowing down canyons igniting grasses left by fallow sugarcane and pineapple fields. Lahaina, at one time the capital of an Indigenous monarchy, decimated by colonialism, capitalism, and climate change.

Since my children were born in the late 1980s, the number of forests burning in western North America has risen with the temperature of the atmosphere. Fewer than ten years ago, the climate crisis was a suspect. Now it's a convicted criminal, as the west turns tinder dry in record-breaking heat, or floods in epic rainstorms.

I used to imagine the apocalypse as one big end of the world, wiped out by meteors or atomic bombs. Now, I wonder if we're witnessing a trickle-down Armageddon, one staggering, home deleting natural disaster at a time.

Yet somehow, the flames missed my brother and sister-in-law's townhouse. The Jesus is Coming Soon church survived one of the deadliest fires in United States history.

And still, the salmon swim upstream, against the current, home.

Two Poems

by Irena Kaçi

Big Sky, Montana

The road was wider than anything, wide enough to match
the sky, and there was so much sky right there over me
with no clouds or nothing for cover.

I used to cower at the lightness of it just over our heads
when we drove at full speed into town, so you could by
your cigarettes, and liniment for your joints, which stiffened

something fierce each time the great big storms rolled in,
huffing and puffing over the expanse of raw land unmitigated
by us, the lonely humans hiding in its canyons. You seemed

so small then, we both must've been two tiny dots, as the hawk flies
moving in short stretches while on foot, faster when traveling inside
that old beater, loving the land when we weren't fearing for our lives.

Isn't that the truth of it? How love and fear hold hands through

*most of it, pushing and tugging on each other when one ends
the other begins? Oh!* I would go back to that beginning,

to those sun blackened bluffs, to that blistering river, to the mountains
stoic and forever snow capped, to the roughage of Indian paintbrush
scraping my ankles with its cat-tongue, I would do it again

and again and again for the love of it, for the fear of it,
to be infinitesimally small at the mercy of the wide road,
and to the roaring sky, a speck of no consequence.

Shoshone River, 2 miles out of the East Gate of Yellowstone

When we rode our horses through it,
it was winding knotted,
muscle all the way down. First the beast
between my legs, then the swollen
river galloping past it,
more alive than I had been
my whole life.

At night I listened to the rage of it.
Thrashing and distant, like it was the stream
of memory beating down the Red Mountains
screaming *I am wounded, I am wild*
I am the lusty west.

We fell in love there. If not with each other,
then with our own young, capricious hearts,
wrenching at maximum capacity,
for each thin drop of oxygen
"Mountain sickness" they called it
that first time I went hiking out there
keeping up, then falling

over, dizzy and gasping the same way
that Yellowstone cutthroat choked
& heaved, after you reeled it in, staring up
at me, and then, at nothing. Its body flopped
indelicately on the boardwalk. I remember
wanting to toss it back into the Shoshone,
where its twisting, would, again, be beautiful.

In Between

by Lawrence Winkler

'We come from the earth, we return to the earth, and in between we garden.'

—Anon

    There's a wonderful joy in turning over the hibernating soil in our raised beds today. I can smell chocolate promise, pungent with Pasternak ozone, with every twist and lift of my pitchfork. This is not one of those life and death activities. It's much more important than that, the big hand on my clock. Cicero said that 'if you have a garden and a library, you have everything you need'. In the garden section of my library, I have a collection of seed catalogues, full of spring anticipation. Gardening begins with the dreams in January, and I dream bigger than emperors.

    The garden is never so good as it will be next year and, every March, when it's still Summer in the light and Winter in the shade, my favorite mail-order seeds start arriving, in coloured envelopes. They find their way into the alphabet of my accordion file box in the shed and, when the right moment in the right week of the right month arrives, claim their own personalized paydirt: Arugula,

Basil ('*Genovese*' and '*Thai*'), Beans (fava '*Superaguadulce*', asparagus yard-long '*Bacello*', bush-green '*Balong*,' and yellow wax '*Rocquencourt*'), Beets ('*Detroit Red*'- although I prefer to plant a mix of '*Chioggia*', '*Cylindrica*' and '*Touchstone Gold*', Robyn's pickling recipe from Nana Pitcon works best with a more rustic approach, Broccoli ('*Comet*'), Cabbage ('*Savoy*' and '*Red*'), Carrots ('*Nantes*' or '*Mokum*'), Corn ('*Bodacious*'), Chard ('*Riccia da Taglio*' and '*Fordhook Giant*'), Ching Chiang, Cilantro, Cucumbers ('*Cool Breeze*'), Dill, Eggplant (Japanese '*Kamo*'), Escarole, Fennel ('*Romanesco*'), Kale ('*Black Tuscan Lacinato*'), Lettuce (too numerous to list, but oh I do love '*Amish Deer Tongue*'), Mesclun, Parsley, Parsnips ('*Gladiator*'), Peas ('*Sugar Snap*'), Radicchio ('*Palla Rossa*'), Radishes ('*French Breakfast*'), Spinach ('*Bloomsdale*', '*Olympia*' and New Zealand), Squash (I usually plant '*Waltham*' butternut and South African '*Gem*', but this year I'm going out on a limb with '*Marina Di Chioggia*'), Sunflowers, Tomatillos, and Zucchini ('*Aristocrat*'- only two plants, only two plants, only two plants...). These are just the seeds. Out in the raised beds, back under the snow (it's the vernal equinox for Chrissake), lie subterranean asparagus spears, strawberry rhizomes, and a buried bulb ballet of Russian garlic and French shallots. Woven through the unique architecture of our raised beds, are small boxes of blueberries, black currants, gooseberries, and rhubarb, and a central arbor of kiwifruit and table grapes ('*Interlaken*' is Robyn's favorite, but we also have '*Flame*' and '*Himrod*', more like a turnip masquerading as a grape).

Just to the left of the raised bed Pagoda are the fruit trees. There is a small orchard of cherries, peaches, pears, plums (our favorite fig jam '*Greengage*,' and an F1 hybrid that reverted to its wild-type the minute it was planted), and apples- including a '*Cox Orange Pippin*' for the vintage Port in the cellar. To the right of the Pagoda is our Berry Run, compleat with golden raspberries, boysenberries, tayberries, and a Himalayan blackberry that got in with a fake ID.

You may have noticed that there are some wonderful things that are not present on this list of planned composted comestibles.

There are no peppers (no Jalapenos—this is unfortunate), no tomatoes (no *'Sungold'*- this is a tragedy), and no potatoes (no *'Yukon Gold'*—this is a catastrophe). The reasons are simple and diverse. First, I just can't grow peppers. Maybe it's the magnesium, maybe it's the lack of a greenhouse. I get them to their shape but not their show business. This is a great source of personal shame (and a retirement project). Second, I used to grow great tomatoes but, in recent years, the blight has been getting straight 'A's' in Natural Selection class. Nothing is more frustrating than watching your heirlooms make it all the way through the summer until, just before their sugar hits escape velocity, it's a Black Death party and you're the caterer. Third, there is one thing that is even more frustrating—the black-tie coon supper club that nails our Yukon Gold every single time. I'll still plant them this year because I'm an eternal optimist with a failing memory.

The Gardener's Dictionary defines a garden as 'one of a vast number of free outdoor restaurants operated by charity-minded amateurs in an effort to provide healthful, balanced meals for insects, birds, and animals.' Among the deer, the rabbits, the raccoons, the mice, the birds, and the slugs, there is hardly room enough for the stuff I'm trying to grow. And in those remaining spaces live the weeds, which differ from non-weeds by how much more easily non weeds come out of the ground when you pull on them. Nature doesn't just abhor a vacuum; she blows. Even my faithful hound, Shiva, has repeatedly been caught red-handed eating grapes on her tiptoes, or standing in the raised bed boxes, pulling carrots out by their tops with her bared teeth, and a flamboyant toss over her shoulder. It's the jungle out there, and there are no property rights. What I can't fathom, is how cabbage butterflies find my cabbages? How does the carrot fly get under the remay to our carrots? John Ruskin once offered: *'the highest reward for a person's toil is not what they get for it, but what they become by it.'* Well, I'm becoming bitter. In 1798, William Wordsworth opined that *'wisdom is oftentimes nearer when we stoop than when we soar'* and even Martha Stewart has acknowledged gardening as a *'humbling experience*' (from the humbling experience herself). My garden may be a thing of beauty, but it is also an unforgiving arena of sex and death.

The word *garden* is from both the Old English *geard* ('fence or enclosure') and *garth* ('yard or a piece of enclosed ground'). The Oxford Dictionary defines it as 'enclosed cultivated ground'. Enclosure is essential to gardening. That is something we thought we did well and, initially, we did.

In the beginning, there was Mike Gogo. In 1993, I rode out Nanaimo Lakes Road, with a plan in my head, and a list in my hand. I needed untreated cedar, enough to build fourteen 12x3 ft. raised beds, two feet high. There would be seven pairs of raised beds, all connected by a long central 'hallway' of as many *torii*. It was Shinto, and *tres chic*. The middle path would be covered with grapes and Kiwifruit- the *Hanging Gardens of Westwood*. On either side of this centre aisle would sit various fruit-bearing shrubs and, external to that, elevated fields of chocolate earth studded emerald green. I had designed it, with some foresight, as a labyrinth of wheelchair friendliness. No cast-iron back with a hinge destiny for me. Mike would charge me a King's ransom for his timber and determined to know beans, I paid it.

Hungry to begin, I was waiting with my belt sander, retro-aural pencil, skill saw, T-square, bubble level, and nuts, bolts, and nails when the truck came down the driveway. A flurry of activity ensued. On one sunny spring day, Robyn and I assembled the pagoda and, a few short weeks later, I had filled the fourteen boxes with topsoil and hope. And grapevines, and Kiwifruit, and rhubarb, and berry bushes. The first year, I couldn't walk outside without tripping over the joy of my own cleverness. Then I realized that I needed more garden space, for those things that didn't like living two feet off the ground. A plot was opened for the three sisters (corn, pole beans, and squash), and some second cousins, the following Spring. We were heading, inexorably, towards the Illyrian goal of a small house and a large (and enlarging) garden, when I realized we had…varmints.

One of my patients was a fisherman, who provided us with enough thick nylon netting to make a true 'enclosure'. Every year required the employment of new defensive strategies, as Mother Nature invited more of her friends to the picnic. Interior netting, mulch, polyester row covers and, this year, slug bait for the first time. This escalation of technology is designed only to retain some portion of the vegetables and fruits of our labour. The ephemeral is of lasting value only if the ravenous appetites in our back yard would back off. Please.

The measure of a great society is when old men plant trees whose shade they know they shall never sit in, and I am, despite all my efforts both ways, getting old. I look to other old men who gardened. There was gardening wisdom in abundance, in the Founding Fathers of our southern neighbour. George Washington hoped to spend the remainder of his days 'in peaceful retirement, making political pursuits yield to the more rational amusement of cultivating the earth'. Thomas Jefferson recognized that 'Though an old man, I am but a young gardener'. And Benjamin Franklin observed that 'man receives a real increase of the seed thrown into the ground, in a kind of continual miracle, wrought by the hand of God in his favor, as reward for his innocent life and his virtuous industry'. Ralph Waldo Emerson nodded to his vegetables, and they nodded back. His garden spade healed all his hurts.

And that's why we garden, isn't it? It's not about the multicoloured chard, or the braised home-grown radicchio, or the size of your beefsteak tomatoes. Too many people spend money they haven't earned, to buy things they don't want, to impress people they don't like. The essential elements of being happy are having something to do, something to love, and something to hope for. I have all three every day as a gardener. That man is richest whose pleasures are the cheapest, and gardening is the purest of these. You can bury a lot of troubles digging in the dirt. Gardening is a medicine with no toxic dose. However, there is no hiding your inadequacies. I can tell who you are by looking at your garden. You must love your garden, even if you don't like it.

We come from the earth, we return to the earth and, in between, we garden. So many seeds and so much space, and so little thyme in between.

'The best place to seek God is in a garden. You can dig for him there.'

—George Bernard Shaw

Land of Spiders

by Kiki Adams

In the lonely
deep of the mountain woods, lost
birds weep towards the hidden
sun, climbing dewy
air to fall with the cold
rain and land
back with the ants in spiderwort,
bloodroot white, infinite chamomile.

From high where the trees
shivering shrink
back, cower facing
the drawn shroud
of fog, echoing
the call comes down
ancient, to lick at eyes,
so many eyes.

Songbirds crawl off crying
that the urge to fall must beckon
into furry darkness, a world
lost, shoved beyond

all reach, buried beneath
soil, skin, and blood
and bone. Burned
with the winter fire that asks, "come

glow and be happy and warm and,"
to one, to ash, to branch,
to the resting Chestnut.
Follow in fevered pursuit
the flies congregating
to receive communion,
the blood and body of the fabled forest
bison. Corpse.

Found and forgotten
cabin of living
wood and ivy clawing
eaten and swallowed in time's ache.
Soil breathes painful
beneath with heavy burdened
back restless old turtle
bathed in poison.

Beasts roaming,
holding, must root,
fighting nail and claw the call to the cold,
but the cabin is closed. Locked. Boarded. Buried.
Inches deep
in dust, windows clouded obscure,
the living left
alone to refuse the fold.

This Land of Spiders, every creeping
speck and reaching
arm, rotted attic to mildewed basement,
there are too many and no space to breathe and no space to live
and

no space to die. The Land of Spiders squished
in fevered desperation for a cleanliness as far
from Godliness as the empty West.
The Land of Spiders sucked dry

by the Land of Lilacs, to a hope and a hole. To a hovel
in shadow of green, the red and blue rise
to mock the birds and the fog who
weep for what was forgotten,
down from the deep mountain thick,
here the Last Spider is left to wander
the hills, dulled, mowed over in worship
of peach blossoms.

Wuthering

by J.J. Carey

after Emily Brontë

my spirit finds a spine
standing by shore
at the tsunami bell
alone at sunrise

recording the waves
so that in another life
I can explain the
terrifying surge

of hoping, asking
for you to recline
between my thighs
so we both might be engulfed

while my fingers
twist and thread your hair
when earth and moon are gone
and suns and universes cease to be.

Dear Jody
A Haibun

by Karen Pierce Gonzalez

Dear Jody

Your Canadian winter keeps circling back around while ours is again on the edge of a relentless drought. Golden State gardens form questions of water: what needs how much and which crops can resist the always-almost-here desert that slowly replaces our once-upon-a-time forests?

Vast lands of timber now wildfire-ashen, we can still remember how green we had it, how pond frogs drank up, but never emptied, shallow pools, how squirrels nestled in plump, protective branches, how yard sprinklers were left on because swimsuits of youth were meant to stay damp.

Changes in weather bend all living joints in unpredictable, sometimes breakable ways. A quirk here, a pain there.

Flower beds wilt in reply. Our skin cracks in places moisture would have not allowed. We live in a tinderbox of our own making and yet, continue to take long showers, wash our cars weekly, and insist on ice cubes in lemonade because we like the way they clink against glass.

California
left coast receding shoreline
waves of ocean dry

Two Poems

by Kushal Poddar

Where You Put Your Junk

Someone mentions an Indian War. Was it Lucille Clifton?
Something about Washoe? I live in the India India. I say. What
does that mean? I shrug.

From the cliff of dreams I leap this night, meet the tribe's
champion dreamer midway in the black water.
He points out at the white shores. Let's cleanse it. He says. I sigh.

I always tell my parents (because in this dream like in most others
I'm in their house) that cleaning means placing the junk of one
place in another.

The Tourism Department Bus Arrives

On one shoulder of a pseudo sphinx
two pigeons sniff the smell of sun
rotting. The bus arrives late.
The card players of the piazza
turns their head because the tourists
expect them to, or perhaps they seek
that face who never returned home.
One of the boxes come apart, split
everything with the cobblestone street,
and as if until that moment no child was born,
the half naked boys scoop out the shattered
bottle of scent, gather the aroma, undergarments;
they gather the lights and the shades;
they take you to the hotel obscura, and
although they may earn from both the poles
it is a good show.

We'll Never Have Paris
A Story

by Zander Lyvers

He unlatches the door to the drab data room, points to a stack of at least ten dusty boxes and slaps me on the back as he says, "Looks like you have your work cut out for ya!"

He starts his exit to the corridor before poking his head back in to say, "Oh, one more thing. Help yourself to the Earthly Delights. On the house for the man of the hour." His artificial smile hangs on his face just a little too long. I can't tell if he's fucking with me or if it's just a glitch.

"Well, I'll leave you to it. And don't forget, bon appetit," he reminded me, nodding at the bowl of energy bars lounging on the table. He knows full well I am not legally allowed to accept them—he's definitely fucking with me.

It's the shittiest room in the complex, like they went out of their way to render this particular room DMV chic. The rest of the environs are modern. Sleek. Exposed brick, floor to ceiling windows with a view of the river, or at least what little river that's left.

They probably had a team-building day where all of the sectors of Earthy Delights came together to smear fingerprints on the door and install antediluvian fluorescent lighting, the kind that gives out migraines like candy. Ripped up the parquetry, only to replace it with pre-scratched, beige vinyl flooring.

They might have made a day of it. Ordered pizza, engaged in ironic trust falls to poke fun at the corniness of corporate-mandated morale boosting. But how do you truly get everyone on board? *Let's decorate a room for the meek man at the agency whose job it is to find flaws in all that we hold sacred!* Brilliant. Everyone hates a bureaucrat.

I get it, I do. But it all seems utterly unnecessary to me. Unless they are living under a rock, they've read about the sad state of regulatory affairs in *The Times*. The OCTA has been defunded and defanged, little by little, from multiple administrations of all political stripes. Death by a thousand cuts.

That's not to say I don't like my job. When I started out, fresh from training, it was meaningful, important work. Sure, it wasn't the most lucrative, but we were making the world a better place, and we were given the proper means to do it, too: the most sophisticated technological gadgets, a budget that saw no bounds, and most importantly, a three-person team. Sam, Sal, and me. The Three Musketeers we called ourselves. All for one and one for all! I never read the book, and neither had Sal, but we think we got the gist. Sam, on the other hand… There wasn't a book she hadn't read. She was the smartest and most striking H.S.S. I'd ever met. But that's irrelevant. The Three Musketeers are kaput. One moved on to greener pastures, while the other was put out to one.

Now it's just me. I like to pretend they're still with me now, slicing open boxes, getting our hands dusty, combing through thousands upon thousands of microchips, trying to find that one lead that could slap a fine on these ethically dubious firms. Gone are the days when we could build a solid case against them. Now it's just audit after audit, and maybe I get through a couple hundred chips of data in a day, if I'm focused. All the same, I still collect a check. Happy to not be made redundant.

    Would it be the end of the world if I took one Earthly Delight? I stare at the packaging, which looks like plastic, but the executives a few floors above would tell me that looks can be deceiving, that they are actually biodegradable. How? I have no clue. My job is spotting data-mined inconsistencies, not conducting science experiments.
    I could use some lithium protein right now. I marvel at the tinsel wrappers—images of naked, dancing hedonists enjoying the fruits of the earth without any concern for where it might lead them by the final panel of the triptych. Could Hieronymous Bosch have ever conceived of a world where fear of the future was more frightening than eternal damnation? Probably not. He was stuck in his time, as I am stuck in mine.
    Quite frankly it's a coup that such whimsical packaging was even approved. This tells me the company must either be in compliance, or possibly well connected.

There are a few petrol stations left on the other side of the tracks. Although technically illegal, they're still relics of a bygone era, and therefore tolerated for historic value.
    Most don't want to be caught dead refueling there. I drove through once for an audit, which was an utter waste of time. "What's the purpose of the OCTA auditing an illegal enterprise?" I asked my superior. They told me that we needed to keep the heat on these places. The laws could change and they might be forced to shut down overnight.
    I remember the faded, washed out plastic images that had been fixed onto the dispensers. Pump number 7 was covered with wildfires, the once vibrant red hues of the flames now rendered pink from the elements. The adjacent diesel pump exhibited a

depiction of empty plastic bottles strewn across an arid plain, which seemed like a more arty, pretentious way to go about it. Did the photographer happen upon these empty plastic bottles in the desert, or did some pathetic H.S.S. interns take their time scattering them around the shoot like Johnny Appleseed?

I recall when the reforms were first passed a century ago, mandating that all emitters graphically display these types of caveats. Sure, it might have given people pause for a second. But after the moment of silence, it was back in the saddle, puttering away, with the A/C and music blaring simultaneously, all thanks to the miracle of internal combustion.

It wasn't long before the mandated graphics became a joke. Someone had scrawled a crude drawing of a hand lighting a cigarette from the top of the wildfire on pump number 7 with a Sharpie. You would be hard pressed to find a warning that had not been defaced.

Of course there were the obligatory penis drawings, too. Kids. Some things never change.

I am hungry.

It makes sense that we're not allowed to accept any gifts from the auditees. It smacks of corruption. But the office is already a shell of its former self, and would it really behoove them to go after the last agent in the Office of Climate Transparency and Accountability? Is it worth it to make a big to-do of taking an energy bar that the company already manufactured anyway? It's not like I'm auditing the aviation industry and they just happen to have some top-shelf Earthly Delights lying around to tempt me, like they're the forbidden fruit.

It would be rude to refuse what's offered to me. Wasteful, even. When I grew up, ages ago, my mother always told me that I should finish my food because someone in the Global South was starving.

These Earthly Delights will probably be chucked in the bin anyway, tainted from their proximity to the sad man from the OCTA.

I'm not sad. Not now, at least. Just hungry. The silver lining to deprivation.

Come to think of it, it should be part of my research to partake. Isn't it limiting to only comb through boxes of petadata just to make sure that the products are ethically sourced and ecologically friendly? It's true, the wrapper propagates that the fine people at Earthly Delights are environmentally conscious, that they are meeting their ESG targets, that for every bar sold, one is donated to the Global South. God knows they need them.

My heart sank when Sam was promoted. She now works with H.H.S.S. lawyers in the Office of Climate Justice. This was a huge deal considering that Sam is an H.S.S., working twice as hard to compete in this globalized economy. She did it though, God bless her. Big raise, corner office, expense account. Sometimes the most egregious greenwashers from our division wind up in her purview, but the irony is that with all their funding and fancy titles, they get even less done than we do. The occasional show trial happens, making headlines briefly before being buried by all the other crazy shit going on in the world. Usually it's an obvious villain and the verdict is a foregone conclusion. Sentencing is typically just a hefty fine. One oligarch is forced to shell out money to an NGO, which happens to be the pet project of the partner of another oligarch.

Even better if the organization specializes in H.S.S. rights. Like they aren't the reason we're in this mess in the first place. But that's the way the world goes 'round. One hand washes the other.

The one-for-one program is a joke. Why not help the H.S.S. who are already here, instead of dumping them on people who lack the proper internal infrastructure to make use of them? Yes, they're hungry, but if they don't have the capacity for processing, what's the point? That's an unpopular opinion, one best left unsaid I suppose.

They charge an arm and a leg for these things. The small bowl of Earthly Delights I'm staring at probably costs as much as my electricity bill. I'm sure Sam can afford these now, even though I don't know what on earth she would do with one. Perhaps she's due for an upgrade. Maybe I'll give her a call when I leave this place. It'd be nice to catch up. We would meet at a bar, and we'd both promise not to talk about work.

It's not like that between Sam and me anyway. At least, I don't think it is.

I still think about that day. The three of us were attending a conference in the city for a long weekend. This was back when we had a budget for travel, creature comforts covered by per diems. Sal had gotten food poisoning from the oysters at the opening night banquet, which pretty much put him out of commission for the remainder of our stay. I felt bad for poor Sal being holed up in his hotel room, but this also meant that Sam and I were inseparable as we traversed the complex, refilling our coffee cups, squinting at name tags, and attending sessions with names like

Data Privacy in Climate Analytics, and *Frauditing: Ozone Loopholes We Can't Risk Ignoring.*

Our elbows were perched on a circular high top outside one of the conference rooms. It was technically a coffee break, an interval that was typically reserved for low-key schmoozing for the driven, and mild flirting for the restless. I was pretty sure our particular collegial coupling was the exception that defied both categories.

"Do you really think we make a difference?" Sam asked, twirling her wooden stirrer into the black swill while the Sucralose dissolved. I considered her question as she folded the three empty packets of sweetener in half, each resembling a paper football. She had such a transformative energy. A modern day alchemist. Give her trash, and you were left with origami.

"Of course. I mean, I think so." I had tried to answer correctly, as if a correct answer existed. Her hazel eyes peered into my soul, conducting an interrogation. My grand inquisitor was trying to find out if we were on the same team.

This intensity made you feel like you were the center of the universe. But of course the opposite was true. I was just another lost soul orbiting Sam's glow, trying not to stare directly.

"Sometimes... I just don't know." Sam said.

I was stifled. Speechless. I finally mustered up the courage to respond.

"We work for a nonprofit. That's something, right?" After I spoke, she exhaled as I deflated. Her eyes scanned down the corridor that opened to the hotel lobby, attempting to concentrate on any nearby ephemera that was less frustrating than my stilted responses. I was failing her test.

"But working for a nonprofit does not magically mean we make a difference." Sam said, giving me another chance to engage in the tête-à-tête.

I took a risk. "Sometimes I worry that we are just symbolic. Like, I don't know… It just feels like what we do is nothing more than a facade. Or a charade. I'm not sure what the right word is." Her eyes snapped back into place, in one swift movement, like a predator on some nature show. I had lied. I preferred to think we

made a difference. But I preferred even more to concur with Sam. I continued, "I worry that our main function is to project accountability, while ultimately just perpetuating the status quo. I think that we're just—"
She gripped my hand. It was at that moment that I noticed she had opted to not wear her wedding ring on that Friday morning.

"Does your room have a view of the skyline or the parking lot?" Sam asked in a hushed tone.

"The parking lot." I answered.

"My room it is, then. I want you to tell me everything."

I rip into the Earthly Delight. I discard the wrapper and it glides down to the grimy vinyl floor. I pull my shirt up and flip open the compartment in the right lower quadrant of my abdomen. Didn't this used to be where H.S.S. kept their appendix? My fingers fiddle with the cartridge, which is hot to the touch. I am far too famished to be graceful. Finally, my raddled Econo-bar clunks down on the table with a thud. Lights flash across the fresh cartridge like a Christmas tree. As the lithium protein shoots a charge throughout my body, I understand why they cost so damn much.

People say that my generation is spoiled. They go on about how Hybrid Homo Sapiens Sapiens have it so much easier than their predecessors. We've read about them in the history books. How they dug their own grave with an excavator before trying to dig their way out with a kids' shovel. They sure talked a lot. The only good they did was design H.H.S.S.

Us, on the other hand… We've offset our dependency on organic nutrients by reducing factory farming by forty percent in the past century. Reforestation efforts have increased threefold now that agribusiness has been incentivized to diversify into the lithium protein racket. We'd do away with most externalities completely if it wasn't for the insatiable hunger of the H.S.S. Sure, some don't have the means to make the transition, but most simply lack the will.

Those made redundant will do what certain people have always done—wallow around in the mud and dig up the elements necessary for us progress-oriented few to preserve the light of consciousness.

Two Poems

by Gurupreet Khalsa

Earth's Vespers, in Gray-lidded Sleep

Cast your eyes over my country
where children came and went

amidst vague fleeting longings,
where men came and went, stomping

boot-clad feet in icy sleeted gusts,
edifices of power long abandoned.

A long time ago the train came and went,
slipping through frozen landscapes;

on naked roads a mother came and went,
rushing to sustenance in bone-pith chill;

a child once dreamed of woe,
weak voice whispering last pleas

as selfish gods demanded sacrifice

of all life, unknowing gods dissatisfied

with their tribute, clouded frustration
obscuring sunlight as it breathed prayers

never heard. Earth called in her debts,
time wiped pallid dreams, haunting humanless

wastes as the sun set on civilization;
when lies reached their storied end,

restless wraiths lurked offstage, chill demons,
sleeping dragons hoarded seasons,

fog and frost picking up uncertain memories
to take to the clouds, floating in sepia shadows.

What need of ancestors when descendants don't exist:
what will respawn in slouched hills

waiting to reunite earth with its ka,
rebirth a nobler world?

Show Me the Way

A dying world, on the brink of ruin,
cows that belch and burp, lost bruins.
Die by fire; pole ice ruptures along
cracked seams; desert-expanding susurrations.
We know what it is, barreling headlong
through unavoidable permutations.

Well, show me the way to the next whisky bar;
oh, don't ask why; oh, don't ask why.

Bees and butterflies diminish;
poisoned water through obfuscations,
plastic pellets choke ocean fish.
We're numbed to climate revelations.
Money, power, spring décor, fashion:
routines of frantic stuff-consumption;
hurry to spend and hoard and cash in,
push ahead, don't stop construction.

Well, show me the way to the next whisky bar;
oh, don't ask why; oh, don't ask why.

Pull levers of change, sudden and hard,
oh, humanity, we are marred:
stop in the moment to make the turn;
we don't need more validation,

creatures among creatures, we churn;
what does it take for conservation?

Well, show me the way to the next whisky bar;
oh, don't ask why; oh, don't ask why.

Bertolt Brecht, Kurt Weill *(Moon of Alabama; Mahagonny, 1927)*
Also, The Doors *(1967)*

(A BOP Poem is a poetic argument of three stanzas, each followed by a refrain taken from a song.)

Shore Birds

by Phyllis Green

Three Poems

By Emery Pearson

I'm Angry at Idaho

I won't ever fold into the bend
of your river, won't ever touch
my tongue to your snow. I'll never
again birth an Idaho child
you'll reject.

Yet I miss your switchbacks,
your elk in the snow, their scraggly coats.
The summer mountain lion,
the charred corpse Ponderosas.
The route that ushered me
to my grandmother, my aunt.
Their bodies are yours now forever.
I've made peace with that. With your ravenous
graves yawning to embrace them.
My grandmother's laugh eternally
haunting your pines.
My aunt's golden hair swept upriver,

billowing among the trout eggs and
rocks made smooth by the immense hands of
mountain men I used to admire.

The Cabin on the South Fork of the Payette River

Sometimes love is a consolation prize.
Sometimes dusk on the back deck,
deer stumbling down the stairs to the river,
the heat giving way to crisp dark air—
sometimes dusk was reason enough.

My bones are part pine.
My blood part burbling creek.
I could sketch you the dartboard and
the Adirondack chairs, each log
of the cabin facade. All the beds.
Mama hawk and her nest in the ponderosa.
Tell you tales of the chipmunk in the woodpile,
the bats that swoop at night.

Rights to the land did not appear
in our final paperwork. I didn't get to say goodbye.
I'll never again put swollen feet into buckets of creek water,
never write next to the river, never watch the moon eat the sun
through paper glasses, never pull rocks from babies' mouths,
never step out of a pickup into high-altitude air,
never see the lines of the year's demarcation,
never call that family mine,
never ever will I ever
get to call that land
mine.

Drought Land

Maybe I'm not wishing
hard enough for rain.
In the backyard the birdbath
dries up again. The rock landscaping
burns my feet when I fill it back up,
when I offer the greedy finches
more wild bird mix.
All day I check my weather app.
The billowing blobs congregate
then disband past the peaks of
the San Bernardino mountains.
To the east, in Las Vegas,
restless monsoons surge through
casino ceilings, rainbow rivulets
joining roulette tables.
The Strip is a river, grabbing
street trash and abandoned loneliness
and heading out to parts unknown.
Everyone speculates about Lake Mead,
the termination of severe drought, as though
we deserve enough moisture
to undo our sins.
Out my own window, to the east,
the dark clouds gather
again. So close but never close
enough. I came here knowing
water would become
an obsession. I fled
from one red state to another
kind of red. A red so dark
it's almost black. I have no grass
but also 100 houseplants,
two children, four pets. Today
I wasted water on a filthy corner of the patio

while gazing at the willful clouds.
The dirt stuck, sunbaked and snide.

Χαρούπι
An Essay

by Ivars Balkits

 It was haroupi day and one that went from early in the a.m. to near sunset. We picked up Dmitri and Zacharias (former agriculture advisor to the government) and took the long road to Melihoris (Honeytown) and above there to Kosta's place. We met him as he was coming down, on his way, I suppose, to arrange for the meat. The lamb (20 kilos worth) was to be cooked antikristo—lit. across the fire—split whole lambs on a wire cage in tiers surrounding a wood fire.
 We had tons of mezes to start: horta (greens) including vlita (wild amaranth) from Kosta's garden, boiled potatoes, pickled cauliflower, offal from the lamb, capers, a kind of soft cheese similar to ksigalo (made from goat milk), wine from Kosta's cellar —on the sweet side, like a tawny port, but not cloying at all. Also from the cellar, kephalotiri (lit. head cheese, but not what we in the US call head cheese), with a chunk bitten out by a mouse. And, of course, all imbued with the aroma of carob syrup cooking down in a special room below the house.
 Kostas has, says Dimitri, about 10,000 trees! From his porch, we see 1,000 of them at best. He has improved the quality of 3,000

of them (through grafting, I guess), says Dimitri, but Dimitri would like him to stay on target and finish with the other 7,000. It is Dimitri's greatest frustration that Cretan producers get distracted, give away too much in feasts with friends, spend their money on tangential projects, and celebrate until they're broke all over and over again. The feast Kostas prepared must have run into 100s of euros. We together bought only 150 euros of last year's syrup (bottled and packaged nicely by Kostas' wife Maria).

The syrup is rendered every year in two large stainless steel kettles (katzania) over a wood-fired furnace—it used to be boiled down over an open fire. Kostas includes the seeds in his process. Commercial syrup makers remove them for sale to pharmaceutical companies – it has health benefits apparently. These big producers also make a much more watery syrup, says Dimitri. They charge about 3 euros a bottle. Kostas must charge 10 euros to make any profit (he generally throw in an extra two bottles on an order of ten). It is not a huge batch compared to what you might see in industrial operations. The process takes a week, Kostas says. The last three days are crucial—the fire must be stoked, the liquid stirred (with a long carved stick that looks like a golf club). He gets help but is mainly awake for those three days.

At about 1 p.m. (13:00) on a Thursday, the syrup is ready to pour into two tanks, one a bit smaller than the other. The taste is mild, between sweet and sour. It looks like cough syrup. Before the proliferation of olive farms, says Dimitri, the island was blanketed with carob trees. In hard times, it was used to make bread (from the inner husk), and so is often looked down upon these days by Greeks as a result of that association. People also likely grew more vegetables for market then and had more citrus (lemons and oranges). Olives are a bit of a monocrop on the island.

Dimitri has been struggling with staying in Crete or finding haven away from Europe altogether, he says, possibly in Lebanon or Armenia. He has also been to Syria. It is that lax attitude of Cretans and maybe all Greeks that bothers him—the political and individual will that is lacking, even indifferent to implementing solutions to social and economic problems. So, he says. He is a bit of a philosopher, besides a booster for sustainable agricultural processes in Crete. He says he would build a house he doesn't like if it fit the needs of the landscape or its purpose. He says our minds, our emotions, deceive us. When you do things right from the beginning to end, your mind is free, you are free.

He has some source of funding, maybe a grant, but does not want to give Kostas or any producer money directly—rather form a cooperative arrangement where everyone knows their role and contracts to perform it. In this plan, Kostas takes care of the trees, another deals with meals for sale (Giorgio the chef who came also), and another deals with the rooms for rent and hospitality. Dimitri has a degree in hospitality from an Egyptian university. Dimitri feels so strongly that his ideas are right that he has sued his father twice to stop him from making bad decisions about the family olive groves and property.

So we had a long day of eating and drinking and talking (I mostly listening). I enjoyed their company though I only caught a word or two in a sentence at best. At times, I got the drift of the conversation—food, gardens, raptors (because Dimitri had been involved in the re-introduction of certain Cretan vultures also). I enjoyed the camaraderie and energy of this group (parea) and they treated me graciously and I did not feel excluded at all. (Kostas offered me a cigar at one point and asked for clarity on my name.) Before we left, Kostas said (through Dimitri) that we could take all the vlita we could gather from the garden. There was a lot of it surrounding the vegetables. Dimitri said, because of the rain, Kostas was not able to plant until about a month ago. Still, it was flourishing as if planted months before. He has a big pile of composted sheep manure to draw on.

As we left the party, Dimitri asked me to look out the window of the car at the ships under one of the one of the large velani (holm oak) trees. I had a tough time seeing any boat there – but there they were, the ships, both black and white, huddled in a mass in the shade, along with the ship dog.

Jonestown and All the Rest of Those Fucking Weirdos

by Lisa Lahey

Loved by all here, where you been man, we been waiting for you. Pure in heart, body, and mind. Drink of me from this dixie cup. Worship the word and find your way. No one will hurt you here. Sitting in tiny, dark room for hours on end. No window, no door. Watching the word, over and over. Neurons and dopamine slashing through your fucked-up brain. Lasers in your eyes and soul. Nikes and track suits. Starman waiting in the sky in his pure white robe and beard. Business suits and flirty fishing. Seek and ye shall find the way to enlightenment and streets carved in golden shit. All this and more. We have what you're looking for. Never stray from us or eat the dirt and the insects. Second thoughts? Thoughts of your own? The shunning the shame. The hacking and hacking into your addlepate brain. We forgive you. Welcome back to the folding jeers and steers of serenity. You fucked up whore. Fingernails melting into shreds of confetti. Whirlpools of piss and tears and all things kaleidoscope. The lows and the heils and the marching death parades. Many more monkeys, many more drums. Off to the ovens, off to the ash. Come hence my child, your putrid diamond dog feces rotting in the earth. Where is the sugarcane?

The sustenance and the sap. We loathe you. We love you. We love you. The flashing lights and the infants crawling on ceilings. The false chair is truth. Listen to us. Listen not to those who would mislead you. I am the truth and the way. Bitches be whores and whores be madonnas. The book according to David to the bible burning when ye seek and ye shall find. Seek out the fornicators and become one with us. No greater love has a man than he would lay down his life for oneness with the Stars. Drink the Flavor Aid and swim through the only way out. Trust in us. Trust in the word. See you on the other side, sucker.

Two Stories

by Lori Litchman

2nd Annual North American Pollination Conference in Response to the Great Bee Disappearance

Let's Roundup everyone, okay?
Welcome to *How to Think Like a Bee*!
We've learned a lot since last year's conference.
This year we are focusing on
SAFETY!
We've just received word that membership is down –
too many folks falling from fruit trees
while using unvetted, extended telescoping pollination poles,
trying to get more bang for the buck.

It's not worth it.
You really need to get up close and *hand* pollinate
each flower if
you want to eat, folks.

We know
it's difficult to pollinate so many flowers,
which is why this year

we are suggesting a team approach!
Why not get the whole family involved!
Tiny fingers can spread pollen more quickly!

Some so-called experts thought pollination
from the sky would help,
but after the government
shot down two of our balloons,
we've had to issue
a pause in unmanned aerial pollination.
Safety first!

Scientists have been working hard to try to bring back the bees,
but there was that freak accident when, somehow,
the DNA from a murder hornet
got mixed in with a honey bee,
and well, let's just have a moment of silence,
for those scientists
lost in the line of duty.

Ok! I'll go over the agenda and then you can
Think Like a Bee
and buzz on over to the talk of your choice.

Our Pollination craft talk this year
will focus solely on coffee.
We need to bring back that morning buzz!

The fruit sessions will be divided into three sections:
trees, shrubs, and vining fruit.
You may only pick one session.

In the afternoon,
we'll shift focus to vegetables.
You definitely don't want to miss
the robo bee demo
in the courtyard

and if you like what you see,
you can head right over to the marketplace
to get your own hive of robotic bees
enhanced with artificial intelligence.
The first 50 people to buzz on over
and buy will get a honey of a discount.

We'll wrap up the day with the keynote,
when author Doug Tallamy will present
his one-man show,
I Told You So.

Make sure you hit up the auction,
sponsored by Monsanto,
where you can bid on a variety
of buzzworthy gadgets and gift baskets
with all the latest tools and mechanisms
to help you get your pollination groove on.

Have a great conference
Don't forget to complete your survey at the end,
and remember to always
Think Like a Bee!

The Administration has accelerated efforts to protect Americans from harmful effects due to Per- and Polyfluoroalkyl Substances (PFAS) exposure

"It was a house-that-Jack-built sequence, in which the large carnivores had eaten the smaller carnivores, that had eaten the herbivores, that had eaten the plankton, that had absorbed the poison from the water." — **Rachel Carson, Silent Spring**

There is limited information available regarding fate.* Exposure to per- and polyfluoroalkyl substances is ubiquitous. Essentially everyone in the United States has PFAs in their blood. Corporations, people just like you and I, have improved our lives so much! All of us citizens united in our pursuit of happiness. *Transport is poorly understood.* We are so grateful for non-stick cookware, oil-resistant food packaging. *Diet was identified as a significant source of human exposure.* We love the guaranteed mess-free carpets our babies crawl on, clothing impervious to all kinds of weather, shoes that will keep our feet dry in the toughest conditions, silky-smooth body lotion, long-lasting lipstick, grease-busting dish soap, stain-fighting laundry detergent. Easy glide dental floss. So much non-stickiness! *Bioaccumulation can occur.* How could modern life exist without the ease of these chemicals? *PFAs may percolate through the soil, underlying ground water. Drinking water. A wide range of health outcomes including reproductive, birth, developmental, behavioral, neurologic, endocrine, immunologic, metabolic, cardiovascular and cancer outcomes.* The one word missing from the 108-page report: forever.

Don't drink the water
Don't eat food or touch the soil
Do not even breathe

* *All italicized text is from The Per- and Polyfluoroalkyl Substances (PFAS) Report from The Executive Office of the President of the United States, March, 2023.*

My Sanctuary

by Claudia Althoen

slow brooks,
whirlpools.
I dance from rock to rock.
a bridge of birds greets me
at the middle of the chasm
where the rush drowns out the sound of everything
post-industrial.

the river's moisture feeds the vegetation climbing the stone wall.
moss lays on the branches like weighted blankets—
dripping paintbrushes that slather me in dopamine.
solitude is a gift that feeds me serotonin,
for my soul doesn't belong in traffic.

in this worn-book yellow sunlight
streaming through the mountain hemlock Jurassic landscape,
I can't hear my breaths, but I can breathe.
the boulders and strewn rocks
are the building blocks of my stream of consciousness,
while my kitchen-timer thoughts are on hiatus
until I return to the modern dystopia.

I'm alone in the best possible way,
and I lay to rest to prolong my stay.
I name every tree 'Peace'
while the warm breeze tucks me in to sleep.

Fallen Fruit

by Didi Aphra

And today we fell with strawberried tongues
with a half blush deep inside our bellies
from soft-bodied fruit
So close to that delectable feeling—
what was it? Happiness?
How amorous the plants are outside
those French windows, bees making love
which you imitated with your thumb
on curved flesh close, so close to the center
I could hear soft hymns from the Muses
echoing in my ear, I could feel the droplets
of dew fall from sighing wildflowers

Natural Extracts

by Skye Rozario Steinhagen

Extract of Almond

May I crack your disdain with
My hardened heart,
So while you recline complacent,
Smirking and snide, you'll
Crunch bits of broken
Teeth along with those
Almonds you keep
Shoving in your
Mouth.

Extract of Pokeberries

Rest is rescinded for the wretched.
Your filth would pervert the dirt,
Make maggots of mulberries.

What difference would you know
From rotting soil, when
All you are is detritus
Straddled by brazen boasting
And flippant falsehoods,
Curdling yourself into spoilt
Dignity curdling,
Until you are sour fumes

Swatted away
By tiny claws scattering
Seeds of fuchsia fruit?

Extract of Nectar

I want to step on your chest
With seven-league boots,
Drench the breath of
Your insipid quips
From smug lungs
Until echoes escape your scathing lips
Swallowed by the air,
And my foot rests
On nothing but
Daffodils.

Extract of Snow-Slush

Oh, you can flourish a cold-red hand
Repeat some insight you've heard but don't know
While my eyes linger on melted ice and dirt
Drizzling from the bottom of your
Knee-propped shoe.
You double-sided demon,
Turn your two-faced head
Back and forth all you like,
But your boots will never
Cleanse themselves of drivel,
Will never light on untouched snow:
Your shoes carry an upturned chin.

And when you've slipped and fallen on
Your own slush and mud,
I will pass by with pocketed gloves and
Regard from afar your
Black-puddle pretense
Stark against the snow.

Extract of Bitter

Bitter is those old khaki men
Who puff themselves in baby powder
After showering and greasing their 'staches before
8am Service Sunday Morning.

That baby powder lingers;
The parishioners know you are
One of those powder-men. And you sit
With other powder-men whose wives are also dead
And shed white dust on the pew and on the hymnal
And on each other and words from mustached lips are blanched.

The only time you're bare of talcum lace
Is below the spray naked and somewhat clean.
Because after the gentle dabbing
Of glib grey towels,
The powder shakes on and for a moment
A blizzard of bitter
exhales snow drifts and flakes—
And then you head off to Mass.

Mariana's Headstone
A Story

by Zach Murphy

The trees are bare enough to see the squirrels' nests. Frederick scratches his gray mustache and squints his weathered eyes, wondering how a creature could rest on such a fragile bed, at such great heights, amidst winds that could carry away a thin branch.

During the spring and summer months, Frederick had spent every morning taking care of his beloved Mariana's gravesite. He'd bring a pair of scissors in his back pocket, get down on his hands and knees, and make sure there wasn't a single blade of grass out of place. A fresh set of daisies, strategically placed in a vase next to the headstone, would add a hint of delicate sun to the roughness of the stormcloud-colored granite.

With winter on the way, Frederick knows it's going to be a lot harder to keep Mariana's headstone clear. The snow doesn't care about the names it covers, and wool gloves just aren't enough to warm up hands that have been cracked for forty years. The daisies will shrivel up quicker, if they don't disappear first.

Frederick stands in front of Mariana's headstone. He envisions himself lying peacefully in the plot next to her. When Frederick and Mariana got married, they'd always hoped that they wouldn't ever be without each other for long. But when each minute feels like an empty lifetime, a day feels like another death.

On the way home, Frederick's walking stick taps against the sidewalk like a ticking clock. His walking stick has seen better days, but so has anything that has traversed the grounds of time. His back seems to hunch more with each step, his frown burrows deeper, and every breath becomes a bigger job when the cold air enters his lungs. The new neighbors whisper to each other from their porch, and Frederick turns away. It's hard to face the world when you're mourning your own.

As Frederick approaches the walkway of his deteriorating Victorian house, he looks up and witnesses a squirrel falling from the birch tree in his front yard. The squirrel lands on the firm soil, pauses for a moment, frozen, then springs up and darts across the street as if nothing happened.

Frederick steps into his home which doesn't feel like home anymore. He hangs up his scarf, caresses the sleeve of Mariana's old coat, and sighs. After making his way up the creaking staircase to his bedroom, Frederick lies down in his bed and stares at the ceiling. A gust of wind rattles the shaky windows. The height of his loneliness makes him feel dizzy. He contemplates whether he'll ever be able to get back up again or not. He closes his eyes and wishes he could be like the squirrels.

Two Poems

by Daniel P. Stokes

A Two Hand Tale

The nest is gone. But not
without a trace. A dirty blotch
within a crook of cable
stands as witness.
Year on year house martins
on a mission swooped deftly
through its funnel. Our co-tenants.
Until the usurpation.
A troop of sparrows,
while the martins wintered,
commandeered it.
An era of dissension had begun.
They screeched day-long at decibels
beyond the legal limit,
coercing, carping, coaxing.
Luring from the nest
is mortal combat. And futile
while the ill-glimpsed world
looms tantamount to hell. At last,

of course, the fledglings flew
and peace descended. And we
had time to steel ourselves
for next year's clutch.

The balcony is quiet this spring
but all the birds are elsewhere.
And when they paint the smudge
and others sit here
will that arc of cable
pass as laxness or be the clue
that cites a two hand tale?
One hand abandoned symmetry
when it encountered
a hub of strife and nurture.
And the other, guided
by demands of hygiene
scoured it from existence
with a spade.

Up a Mule Track

Once you cross the bridge at Aljarroba
you're on a spiral – steel barriers
fronting scarps I wouldn't glance down –
to Sayalonga. She awaited,
an Austrian lady, late sixties,
(undergoing, said the agent, a divorce),
motioned "House is farther". Led us
to a mule track. Tailed her, thankful
the bloody car was rented,
a thumping two miles up.

Past the gate and three slack bitches yelping.
It was showtime. The pool voila,
patio shazam and boom-boom through lawn
And fruit trees. Your beams sufficed.
She nudged us to a vantage
and, pointing, played her trump, her ace
her banker all at once. A rush
of brightness rose and tugged us to it.
The seashore flared, two thousand feet below
at Torrox Costa.

Indoors was proudly bared
to our inspection. The kitchen pinged.
Her bedroom flounced. Telescope
and sketch-board in the spare.
Red satin cushions shared the lounge
with laptop, phone and coffee-table glossies.

And that was it. The tour was over, till
you, emerging from a time-warp, praised
monograms embroidered on the sheets.
"Up here you have much time, you can do

Much things," she smiled, then let us see
her sadden. We sauntered to the garden
once again.
             "Up there, he too is British."
"We're not British."
.   "He is very good man.
Very good friend. Over there is German.
There is Dane. And there, another one
is British."
             "And there?" I pointed,
a rawly finished villa just below.
"They don't live here. They are young, only
forty-five. They must work," she shrugged.
"They come there maybe four times in year,
always two weeks."
             Returning, seeking
something non-committal,
             "The furniture?"
I muttered. Blankness .
             "It's included?"
"No. I take to new house I must get."
I, signifying nothing, nodded.
Turning, she asked as afterthought
could she, if we took the house, have
time to move,
             "One half- year maybe?"
" No rush," I said, reversing
through the gateway, full certain
the pretty house she grieved for
and this paradise of sun and views
and cosy resignation
was someone else's dream.

A Few Final Words

by Emily Bright

Growing Season

That summer, I would hoist the baby
on my hip to tend the garden,
watch with one eye as her sister
picked bouquets of dandelions. Weeds
were my task, plus the planting, plus
the rest of it. Such short growing season
in the north, yet everything that summer
kept busting into bloom, zucchini leafing
almost fast enough to see, the baby
with her first-tooth squalls, and did I
mention the weeds? It was an empire of rabbits,
even with the wire fence. I planted only
what they would not eat, and still

I sent them hopping every time I entered,
scattering my apologies behind them
while the children, startled, clapped.
There was so little time before the baby
started fussing or her sister grew too hot and

sought attention. At least by mid-July
we had the hide-and-seek of orange blossoms,
then ballooning squash, sometimes two inches
in a day. The only way she'd each zucchini
was freshly discovered. That summer,
habit set, eyes and expectations tuned,
we tromped barefoot to the garden
to uncover what was new each morning.

Warming

Thick gray woolen sky today
The scarlet maple
Sprouts red tufts like pompoms
Six weeks too early

This year we skipped winter
Almost entirely
I want to say, hush now
Go back to sleep

But I know her path is set
Her leaves will come blazing
Out into the world
To see what it has to offer

A Parting

 You look around at it all, at the terrain you've traversed, at the bugs and the brambles and the glistering dawn coming up over the mountain peaks. You reach a hand into your pack, touch the fruit and nuts you've gathered as you've walked through this world, seeking and looking and listening.

 This feeling in your head, streams of life and death sustaining each other in calligraphic swirls that pulse through a cosmological compost, must be what John Denver had meant by "Colorado Rocky Mountain high". The breeze around you is gilded by the buzzing of Hymenoptera—honey bees and sweat bees and hornets, who are, after all, pollinators too. Your fingers close over the fuzz of a kiwi and you reach for your dagger, shaving the skin from the fruit in a long ribbon.

 As you lick the last of the nectar from your fingers, your host's voice floats to you on the breeze. "You're enjoying yourself, then?"

 You've seen her now and then as you've spent your time (hours? days? months?) here, walked with her sometimes amongst the trees, heard her sing and heard her stories as well of the land she tends. She spoke of golden fleece and golden apples and golden touches, but never of herself. She comes to you now up the hill where you're standing, under a bonsai tree that twists up like a spiral staircase.

 "Would you join me again?" she asks, and you realize with a jolt that for the first time, she does not carry her basket. Instead she holds out to you a glass bottle, the first thing she has offered that you have not had to harvest yourself. The part of yourself that is still in the woods steels for the trick, for the trap, for the startling realization that it was folly to stray from the looping path after all—but the part of you that has been sustained by the gifts here

know that this is no glamor, no trick. This is the real deal, raw and rare and red as well as green.

You take the bottle and put it to your lips, tasting cold lager, bitter and balanced and just the slightest bit sour. She raises a bottle of her own, the clink of the glass against yours an industrial thunderclap in this untouched World.

"No World is untouched," she says. "Worlds touch themselves, growing hands from up out of their own clay and pottering out their existence. Gravity. Acidity. Pressure. Even in the deadest deserts on the furthest coldest rocks, the wind caresses." She sips her beer and offers you her free arm, the one from which the basket usually hangs.

"Walk with me while we drink."

You feel you should be nervous making contact, but your arm slips easily into hers and you walk hip-to-hip, beginning down the hill. Neither of you speak as you circle broad pockets of poppies, as you pass a poplar full of crows and jays, as you notice a single red carnation standing alone at the center of a fairy ring of amanita muscaria. Your beer never seems to be less than half-full.

The music of several dozen chirping frogs fills the silence as you walk with her to a lily-laden pond, black water catching the first fingers of sunrise in its hematite ripples. You look for the frogs but don't see them, only hear their ringing chorus. Her arm falls out of your grasp as she plops down on a log, patting the moss next to her. "Sit."

You look down at the bottle in your hands, feeling the moisture beneath you, then at the pond, which almost seems too rectangular, too much like a doorway in its geometry, to be purely natural.

Her hand touches your shoulder, and for the first time, you find it hard to look when up to this point looking has been like tasting grenadine. Now the concentration feels too sweet, but out of respect you force your eyes to look once more. You tense under her touch, dreading the coming eviction. You have known since you first touched this soil that the touch would be temporary.

"I'm afraid the time has come for goodbye," she says, and you close your eyes. She has seen you weep as you explored her grounds, and you will not be a glutton, will not shed tears now over being asked to take only what you need, to not drink too much, to not overstay your welcome.

Her hand slips down your arm and her fingers between yours, squeezing. "I'm sorry that it has to be this way. But there are rules to this sort of thing, you understand." She looks at the horizon line, layers of silted sky that look like a slice of canyon.

"If you decide to stay for a while, take good care of her." She tips her bottle at you. "No littering."

You blink, only half-understanding until she stands back up to her full height. "Everything in its time, and mine is almost here," she says. "I really have appreciated the company, but I have other duties to see to."

Still she wears her odd, sad smile, as if it is not an expression but something essential in her nature, a lemon twist of tragedy fundamental to whatever it is she is. For a moment you're sure it's a cruel trick, that this reversal is the trap you dreaded, but you see the truth in her face, if you could ever really doubt her voice.

"Your time is your own, and you're welcome to spend as much of it here as you like," she says. "Wonder. Wander. Like I told you at the top—it's yours. All of it. Every drop of dew, every pebble and boulder, every avocado pit and every glistening peyote. They're all yours, and you theirs. Don't forget the last bit, please. It can be an easy slip. Like landing on a banana peel and toasting a continent with copper." For the first time you see her smile falter, as if this is almost enough to break it when the sorrow of her own situation was not enough. "As for me... well, the world of the gods was not made for the gods to live in. Those who live in the world and make idols of themselves at once all too often make themselves pests as well. Be a weed, spread, but try not to be noxious." She kisses the top of your head and then lightly brushes your hair, like a squirrel burying a nut. "You may stay as long as you wish. Somewhere there's a shed and some tools, and there's plenty of wood to build a house if you desire, if you pick the right trees to talk to. Or if you want to return, the door is still around here somewhere, or..." She gestures towards the mountains. "The World is wide as well as many-faced."

You run your thumb along the sculpted sand of the bottle's lip, wondering when invention becomes incursion and invasion, almost wishing you had been exiled, had been relieved of your post. The surface of the pond trembles, and then you hear it, a low whistle, the call of a long dark train coming up a long dark way from way, way down below. It singes through the soil, a grasping, grabbing sound, turning coal into propulsion and propulsion into diamonds.

The bloated earth by the water's edge splits and furrows, a mechanical lift rising out from the ground in a halo of silky black smoke, groaning, moaning, grasping. The door dings as it opens, a sooty, hungry maw rolling out the reddest of carpets. With a whoop and a howl she turns her bottle upside down, *sending it cascading and sudsing through the mud, writing her name one last time before she bounds across the marshy ground and lands hard on the lift, staining the shining steel with night dark mud. The gate dings again, and you watch as she leaves to attend to her obligations elsewhere, leaving you to your own devices.*

As the dirt settles the only trace of her exit is a sliver of sulfur amongst the perfume, and all of a sudden, your bottle is half-empty.

You finish it slowly, watching as the dawn continues its creep up over the walls of the World—wondering what you will seek, hoping you'll know the direction when you find it, like the wind once knew how to make the primordial earth just the right temperature for eukaryotic life. Your motions ape hers near the last drop, giving it back, tipping the house for its hospitality. When it is empty, you wrap the bottle carefully with cloth and tuck it amongst your vittles, safe from doing harm, capable, perhaps, of recycling, of becoming some as of yet unknown good.

Contributor Bios

Tricia Knoll ("What I Ask of the Future") is a Vermont poet who often writes eco-poetry. She grows many types of gardens—for butterflies, for vegetables, for flowers and pollinators. One of her recent collections, *One Bent Twig, is full of poetry about trees she has planted, loved, or worries about due to climate chaos.*
—Twitter: @triciaknoll.wind
—Website: triciaknoll.com

Kostandi (she/her, "Nature Lover", "In the Afternoon") is a lover of books and cats. She found her passion for writing as a young child. Growing up with a speech impediment, she found writing the best way to express herself. Currently, she resides in Kalamazoo with her husband and kitties. She has been published through Western Michigan University and Sigma Tau Delta.

Paul Hilding ("Before the Storm") is a retired lawyer, living in Idaho. He has recently tried his hand at creative writing and has had one short story published in *After Dinner Conversations*. He just completed the first draft of a "Cli-Fi" novel and is currently trying to find the motivation to do some badly needed revisions.

Claire Thyne ("proliferation", "Meeting the Forest Floor") is an MA in English candidate at Brock University. Her research interests include queer ecology, speculative fiction, fairy tales and fables, and picturebooks. Claire recently found a passion for interrupting the traditional, often stifling academic writing process by experimenting with creative forms. Claire enjoys spending time with her dog, Milo; watching anime; participating in social justice initiatives; resting; and all things mushroom-related. Among other miscellaneous things, Claire is working towards building a better relationship with bugs.

Keiraj M. Gillis ("Brassy") is a gothic and spiritual poet whose works explore the mind and esoteric. His poetry collections include *St. Sagittarius*, *The Gentleman Vagrant*, and *Handsome for One More Day*, which are projects that have allowed him to document his spiritual journey. He enjoys his work as a publisher and spends his time immersing himself in the culture of the American South and Southwest.

Diane Funston ("Taken in Small Measure", "Climate Changes", and "The Secrets They Told Me", and "On this day") recent Poet-in-Residence for Yuba Sutter Arts and Culture for two years, created online "Poetry Square" bringing together poets worldwide. She has been published in *F(r)iction*, *Lake Affect Magazine*, *Synkronicity* and *Still Points Quarterly* among many others. Her chapbook *Over the Falls* was published by Foothills Publishing. Diane is
on the spectrum of neurodiversity and her personality type, INFJ, is the rarest in the world.

J. J. Stewart ("False Spring") is the current pen name of a California-based writer. Under this name, they have appeared in the *Abergavenny Small Press*, *Nat1's Audience Askew*, and the upcoming edition of *Christmas Spirits* magazine. Under other pen names, they have been published in *Blink Ink*, *Boozeleague*, *Calm.com* and others.

William Doreski ("Living Rough", "Bogus Portrait of Shakespeare", and "Two Cherubs") lives in Peterborough, New Hampshire. He has taught at several colleges and universities. His most recent book of poetry is *Venus, Jupiter* (2023). His essays, poetry, fiction, and reviews have appeared in various journals

Kathy Pon ("Orchard Language") earned her doctorate in education, but in retirement has turned to her life-long passion for reading and writing poetry. Her husband is a third generation farmer, and they live on an almond orchard in rural California.

Their poems have been featured in *The Write Launch, The Orchards Poetry Journal, Eunoia Review, Penumbra* and *Passengers Journal*.

Cassandra Caverhill ("While Biking With My Father Who Has Alzheimer's," "Direction," and "It's June") is the author of the chapbook *Mayflies (Finishing Line Press, 2020)*. Her work has appeared internationally in journals across the US and Canada, most recently in The *Coalition, Pagination,* and *Short* Reads. Cassandra is a graduate of Bowling Green State University's MFA program in poetry, and she teaches creative writing in her hometown of Windsor, Ontario. Learn more at cassandracaverhill.com.

Robert Witmer has for the past 45 years lived in Tokyo, Japan, where he served as a Professor of English at Sophia University until his retirement in 2022. He still teaches a course in Creative Writing at the Japan branch of Temple University. His poems have appeared in many journals and anthologies. He has also published two books of poetry: *Finding a Way* (2016) and *Serendipity* (2023). Besides these original works, he served as the lead editor for a series of translations of contemporary Japanese plays, *Half a Century of Japanese Theater*.

Svetlana Litvinchuk ("Just Say the Word", "Red Planet", "Sorry for the Interruption", "Tornado Season", "Let What May", "I am Not Separate", "Grinding Down a Tree") is a permaculture farmer who holds BAs from the University of New Mexico. She is the author of a poetry chapbook, *Only a Season (Bottlecap Features*, 2024). Her work has appeared or is forthcoming in *Sky Island Journal, Littoral Magazine, ONE ART, Longhouse Press,* and elsewhere. Originally from Kyiv, Ukraine, she now lives with her husband and daughter on their organic farm in the Arkansas Ozarks.

John Winfield Hoppin (he/him, "Poem on a Hummingbird Sage") emerging from intersecting social, environmental and physical catastrophe is an artist and poet living and working in San

Leandro, California. In 2001, he received his bachelor's degree from the California College of Arts and Crafts in Film, Video and Performance. Creator of Hoppin Hot Sauce, "The Best Sauce In The World, I'm Tellin' Ya," Hoppin has multiple sclerosis and hosts the *What's The Matter With Me? Podcast* to find support and explore disability theory. Chernobyl happened on his seventh birthday.

Katie Bausler ("JESUS COMING SOON") is a writer and podcaster. Published written work includes columns, poems and essays in publications including the *Alaska Beacon, Alaska Dispatch, Edible Alaska, Wildheart, Stoneboat, Tidal Echoes, Cape, Cirque,* and *Insider*. She also hosts and produces the *49 Writers Active Voice* podcast with writers and artists on these pivotal times, writes a newsletter focused on alpine skiing, and is a volunteer public radio DJ and host. She previously worked as a public media host, reporter and producer. She is working on a collection of essays and poems, working title: *Live Like You're Dying*. Katie and her husband Karl live near their children and grandchildren on Douglas Island along a saltwater alpine fjord in Juneau, Alaska's capital.
—Instagram: @katiebradio
—Facebook: Katie Bausler

Irena Kaçi ("Big Sky, Montana" and "Shoshone River, Two Miles out of the East Gate of Yellowstone") and is a poet and writer living in Worcester, MA with their spouse and children.
—Instagram: @redbicycleblues

Lawrence Winkler ("In Between") is a retired physician, traveler, and natural philosopher. His métier has morphed from medicine to manuscript. He lives with Robyn on Vancouver Island and in New Zealand, tending their gardens and vineyards, and dreams. His writings have previously been published in *The Montreal Review*. Some of his other work can be found online at lawrencewinkler.com.

Kiki Adams ("Land of Spiders") is fascinated with the structure and patterns of language. She explores these patterns in her work as a linguist, but has also been writing poetry since she was a child. Her poetry often plays with themes of displacement, isolation, and the liminal. Kiki studied linguistics and poetry in her hometown at the University of Texas at Austin, and now lives in Montreal, Canada. When not working with words, she can be found watching birds or practicing aerial circus gymnastics.

J.J. Carey ("Wuthering") is a poet and writer surviving late capitalism with the support of a tiny circle of fellow state enemies in Leeds, UK. They write about the urgency & beauty of collective liberation. You can find more of their work on instagram at @vinesthruconrete.

Karen Pierce Gonzalez ("Dear Jody") is an artist and award-winning writer whose work has appeared in numerous publications and podcasts. Her chapbooks: *Coyote in the Basket of My Ribs* (Kelsay Books), *True North* and *Sightings from a Star Wheel* (Origami Poems Project). Forthcoming: *Down River with Li Po* (Black Cat Poetry Press), and *Moon Kissed* (with a North American publisher). 50+ of her images have appeared in various print and online publications; six of them as cover art.

Kushal Poddar ("Where You Put Your Junk" and "The Tourism Department Bus Arrives") is the author of *Postmarked Quarantine* and *How To Burn Memories Using a Pocket Torch* has nine books to his credit. He is a journalist, father of a four-year-old, illustrator, and an editor. His works have been translated into twelve languages and published across the globe.
—Twitter @Kushalpoe

Zander Lyvers ("We'll Never Have Paris") currently lives in Madrid, where they teach Social Studies. Their primary inspirations are George Saunders, Eleanor Catton, and Paul Beatty.

Gurupreet K. Khalsa ("Earth's Vespers, in Grey-Lidded Sleep" and "Show Me the Way") is a current resident of Alabama who considers connections, space, time, reality, illusion, and possibility. She holds a Ph.D. in Instructional Design and is a part time instructor in graduate education programs. Her work has appeared in multiple journals; many poems have received awards.

Phyllis Green ("Shore Birds") is an author, playwright, and artist.

Emery Pearson ("I'm Angry at Idaho", "The Cabin on the South Fork of the Payette River", and "Drought Land") is a writer living in southern California. She has an MA in rhetoric and composition from Boise State University and an MFA in creative writing from Antioch University. Her work has appeared in *Assay, Punctuate, Jersey Devil Press*, and elsewhere. Find her at emerypearson.com and on Instagram @hello_emery.

Ivars Balkits ("Χαρούπι") is a dual-citizen of Latvia and the USA since 2016 who lives part of the year in Ohio but mostly in a small mountain village in Crete, Greece. His poems and prose have been most recently published by *Mercurius Magazine, Sortes, Vernacular, Random Sample Review, Pnyx,* and *Punt Volat.* He is a recipient of two Individual Excellence Awards from the Ohio Arts Council, for poetry in 1999 and creative nonfiction in 2014.

Lisa Lahey's ("Jonestown and All the Rest of those Fucking Weirdos") short stories and poems have been published in *34th Parallel Magazine, Spaceports and Spidersilk, Altered Reality Magazine, Why Vandalism? Suddenly, and Without Warning, Five on the Fifth, Ariel Chart Magazine, Vita Poetica,* and *Literally Stories*. Her work has also been accepted by *Same Faces Collection, Piker Press, Epater, Bindweed, The Pink Hydra, Adelaide Literary Magazine,* and *Creepy Podcast.*

Lori Litchman ("The Administration has accelerated efforts to protect Americans from harmful effects due to Per-and Polyfluoroalkyl Substances (PFAS) exposure", "2nd Annual North

American Pollination Conference in Response to the Great Bee Disappearance") is an outdoor writer based in Philadelphia. She is the author of *60 Hikes within 60 Miles: Philadelphia*. She holds an MFA in creative nonfiction from Goucher College.

Claudia Althoen ("My Sanctuary") is rooted in the vibrant cultures of Edmonton, AB, and Minneapolis, MN and finds solace and inspiration in the written word. For her, writing is not just a form of expression but a way to navigate and understand the complexities of the world and the human experience. She has been published in *The Ekphrastic Review*.

Didi Aphra ("Fallen Fruit") is either daydreaming, frolicking in the woods, or she is in her lair concocting wondrous word potions. Didi has previously been published in *COSY MAG* and her work primarily explores mythology, belonging, and sexuality. Didi is currently working on a short story inspired by the ancient cults of Aphrodite. Her debut poetry collection, *Tender Philosophia*, is set to release in 2025.
—Instagram: @didi.aphra

Skye Rozario Steinhagen ("Natural Extracts") is a latina poet from Iowa, who earned her MA in Theological Studies from the Harvard Divinity School in 2021. Her passions intertwine among English, Humanities, the Study of Religion, and Creative Writing. Her writing has appeared in The Green Shoe Sanctuary, Humana Obscura, and elsewhere. You can find more about her at skyeroze.wixsite.com/skyerozario.

Zach Keali'i Murphy ("Mariana's Headstone") is a Hawaii-born writer with a background in cinema. His stories appear in *Reed Magazine, Maudlin House, The Coachella Review, Raritan Quarterly, Another Chicago Magazine, Flash Frog, Grub Street*, and more. He has published the chapbooks *Tiny Universes* (Selcouth Station Press) and *If We Keep Moving* (Ghost City Press). He lives with his wonderful wife, Kelly, in St. Paul, Minnesota.

Daniel P. Stokes ("A Two Hand Tale", "Up a Mule Track") has published poetry widely in literary magazines in Ireland, Britain, the U.S.A. and Canada, and has won several poetry prizes. He has written three stage plays which have been professionally produced in Dublin, London and at the Edinburgh Festival.

Emily Bright ("Growing Season", Warming") has a collection, *Fierce Delight: Poems of Early Motherhood,* coming out from North Star Press next spring. She holds an MFA in poetry from the University of Minnesota and a BA from Williams College. Her individual poems have been published widely in such journals as *The Pedestal Magazine* and *America*.
—Website: www.emilykbright.com
—Twitter: @emilykbright.

Jasper Martin (cover art, a.k.a. Jupiter Menagerie) is a queer, neurodivergent artist living in northeast Ohio. When they're not busy illustrating, designing, or organizing some sort of creative endeavor, they're likely spending time in their garden, with their pets and loved ones, or playing video games.

Ju Collins (guest editor) is an aspiring writer and editor based out of the rain, drizzle and fog capital of Newfoundland and Labrador. They have been previously published in DPL's queer horror anthology, *Trans Rites*. Their preference is in writing horror stories, a perfect pairing with their M.Sc in Environmental Policy & Management (Energy & Sustainability).
Instagram: @ju424

Fritz Dries (editor) is a poet and laborer from Ontario. He has published three collections, *Harvest Dogs, Four Seconds* and *Bury Your Teeth in the Yard.*
Twitter: @howlingmoth

v.f. thompson (editor) is just compost in training. She can be found clowning around Kalamazoo, MI.
Twitter: @VF_Thompson
Instagram: @v.f.thompson

Acknowledgements

This book would not be possible without those who supported our previous anthology, *Trans Rites: An Anthology of Genderfucked Horror*, as well as those who have supported us on Patreon or in other forms over these first few years of development.

Poppet, Charlotte Gremel, Antoinette Del Rae, Tricia Flowers, Alex Riley, jessica gilbert, Makarenna Binimelis, Ju Collins, Lexx Ambrose, Heidi Luby, Heather Green, Carter Cesareo, charlearning, kungfuturtle, Mal Harrigan, Heather Tobin, blaneyma1, Vanessa Hillier, Emily Brent, Dion Power, Alison Power, kpuddister3, Ally Thoden, Kellan McCormick, Leilani Roser, Kirsten Craig, Courtney Bennett, Lisa Campbell, Bridgit Shebib, Emily Shebib, Solaris O'Dell, Christina White, Oma Meade, Soundararajan Varathappan, Mitchell Noel, Kayla Dober, Josh Søn Af Mørten, Stephen Maley, Amanda Hickman, Meagan Thompson, Kel Craghead, Katelyn Hayden, Vivek Subramanian, Nikki Kumar, Karan Kaul, Beth Stephens, Beth Stephens, Dan S, Kat Jepson, Rowan Wright, kale woods, Mathias Lobban, Wendy Boden, Krista Lindemann, Amanda Keith, dasue123, Rebecca Mead, Sy Mander, skandarks, Ria Doolan, Shanmugam Krishnasamy, Sephi Coleman-Tunney, Jay J, Asherah Sussmeine, Kelly Thompson, Books & Mortar, Tom and Sandra Jatkowski, Jennifer Fargo, Kelly Campbell, Dean & Michigan News Agency, Juniper Books, Alexander Gray & WMU Student Center staff!

In addition, we would like to heartily thank both **Dr. Stephen Covell & The Department of World Religions & Cultures** at Western Michigan University for acting as advisor to this project and the **WMU Office for Sustainability** for their patronage of this book through the Student Sustainability Fund, which would not have been possible in this form without their gracious contribution.

Enjoy this book? Consider subscribing to future issues and other Dionysian Public Library projects at patreon.com/dionysianpubliclibrary.

Don't want to subscribe, but want to be kept in the loop? Visit us at dionysianpubliclibrary.com or follow us at

Twitter: @DPublicLibrary
Facebook: DioPublicLibrary
Tumblr: dionysianpubliclibrary
Insta: @dionysianpubliclibrary

You can also find our editors on Twitter at and @VF_Thompson and on Insta at and.

Interested in contributing to future projects?
Take a gander at dionysianpubliclibrary.com/submissions